Rabbits for Food and Profit

Rabbits for Food and Profit

Edited by Lee Schwanz

Cover design Al Jacobs
Book design Patti Penegor
Recipe Editor Kathleen Schwanz

Library of Congress
Catalog Card No.: 81-71695
Revised 1986 Edition

P.O. Box 624, Brookfield, WI 53008-0624

Our thanks to these sources

The editor expresses thanks to the federal, state and private organizations who co-operated in the editing of this book. Information has been excerpted and quoted from the following sources.

Agriculture Canada—Raising Rabbits

American Rabbit Breeders Association—A Practical Beginning to Successful Rabbit Raising

The University of Arizona—Rabbit Raising in Arizona

University of Arkansas—Arkansas Farm Research

University of California—Rabbit Handbook

Carnation Company, Milling Division—Rabbits

University of Minnesota—Your 4-H Rabbit Project

Oregon State University—Journal of Applied Rabbit Research

The Ohio State University—Rabbit Project

Pacific Northwest Cooperative Extension—Your 4-H Rabbit Project

Pel-Freez Publications—The Pel-Freez Plan

The Pennsylvania State University—Correspondence Course, Rabbit Production

Ralston Purina—Rabbit Book

Rutgers University—Proceedings, Rabbit Disease and Management Conferences

Texas A & M University—Rabbit Manual

U.S. Department of Agriculture—Commercial Rabbit Raising, Selecting and Raising Rabbits, Domestic Rabbits, Diseases and Parasites

University of Wisconsin—Wisconsin 4-H Rabbit Project

Special thanks to the rabbitry owners who sent us marketing ideas and recipes for tasty rabbit dishes.

Table of Contents

Photo and Art Credits

American Rabbit Breeders Association 27,31,36,37,38
Circle K Industries 50,51,53,54,55,57,58,60,78
Carnation Company 86
Lynn Cox 106,107
University of Minnesota 23,25,27,29,61,64,65
Oregon State University 18,19,40,41,52,60,62,78,128
Pel-Freez Publications 34,43,71,115,123
Rabbits Magazine 11,17,23,31,42,46,81,84,98,101,112,145
H.L. Schwanz 96
U.S. Department of Agriculture 74,93,105,125,126
Youngstown Vindicator 99

Introduction

The rabbit is a remarkable animal

"Three does and one buck rabbit can produce more meat in a year than a beef cow and can supply the protein requirements of a family of four."
—Dr. Peter Cheeke, Oregon State University

Dr. Cheeke's statement is the key to the growing popularity of rabbits as a source of nutritious meat for Americans. The prolific, productive rabbit offers an opportunity to grow food for the family and for profit through the sale of meat.

Let's take a look at those rabbit figures. If each doe raises a litter of eight five times per year, she will produce 40 offspring. At four pounds each at maturity, that's 160 pounds. Three does could produce 480 pounds of meat. By comparison, the average beef cow weans a 450 pound calf each year. You can see why rabbits have so much potential as a back yard project for the family or as an enterprise run for part-time or full-time profit.

Mother nature made the rabbit a vital link in the food chain. In the wild, the rabbit is a major source of food for hundreds of animals and birds of prey. The rabbit's key to survival is the ability to reproduce at a remarkable rate.

Most females reach maturity within six months. Then, with a gestation period of 32 days, they produce a litter that averages eight bunnies. In about eight weeks these rabbits are weaned and ready for slaughter at four pounds each.

Commercial breeders are stepping up the cycle with early breeding and weaning. Some are getting as many as seven litters per year. This could mean 224 pounds of meat per year from each doe! That's why breeders are so excited about the potential of commercial rabbit population.

A popular meat in Europe

Rabbit is a popular meat in Europe and in Asia. Spaniards eat an average of 6.2 pounds of rabbit each year. Per capita consumption in Italy is 5.1 pounds and in France it is 3.5 pounds. Here in the United States our total use of rabbit meat is estimated at 25 to 30 million pounds per year. That's only a tenth of a pound per person!

Why are Americans so far behind other countries? Part of the reason is that producers have not been able to supply the market on a consistent basis. Production is fragmented and has seasonal ups and downs. We are now just beginning to penetrate supermarket meat counters where most Americans make their food buying decisions.

Another problem is the image of the rabbit. Children grow up reading about Bugs Bunny, Peter Rabbit, Uncle Wiggly and Watership Down. Children have a special place in their hearts for the rabbit. It frankly is hard to put a childhood image on the dinner plate. At the same time we don't even think about the relative of soft-eyed Elsie the Cow who is in our hamburger.

Dr. Cheeke calls the rabbit "a protein factory." He says, "they are capable of providing more meat per unit of grain and forage than any other domestic animal."

The quality of rabbit meat is superior in many respects. It's a light, sweet meat that tastes a lot like chicken. Rabbit is higher in protein than other meats and lower in the cholestrol, which is a health worry to many people.

Here's a U.S. Department of Agriculture Comparison:

	Protein	Percentage of Fat	Moisture	Calories per Pound
Rabbit	20.8	10.2	27.9	795
Chicken	20.0	11.0	67.6	810
Veal	18.8	14.0	66.0	910
Turkey	20.1	22.2	58.3	1190
Beef	16.3	28.0	55.0	1440
Lamb	15.7	27.7	55.8	1420
Pork	11.9	45.0	42.0	2050

As you can see from the above table, rabbit meat is higher in protein, lower in fat, moisture and calories than any of the other meats commonly eaten by Americans.

Paula Howat, home economics professor at Louisiana State University, says research shows that a 3 oz. serving of baked domestic rabbit offers 31 grams of protein compared with 27 grams for a comparable serving of chicken. It contains only 177 calories compared with 286 calories in a similar serving of ground chuck beef. A 3 oz. serving of stewed rabbit contains 10 grams of total fat compared with 12.5 grams for stewed chicken.

A nutritional and health and diet study found that rabbit has the lowest cholestrol count of any meat. Rabbit had a count of 50, dark chicken 60, light chicken 90, pork spare ribs 105, lean beef 95 and fatty beef and veal 125 each.

Rabbits also are much in demand as laboratory animals for use in major company research. They are a producer of fur for many uses. There is a large group of breeders who like to raise and show rabbits. And then there are many families with rabbits as pets. All in all, we are producing around 20 million rabbits per year, here in the United States.

A fine backyard animal

The rabbit has many advantages as a backyard animal. First of all, you get a lot of production for your money. The rabbit is gentle, makes little noise and is clean. It adapts well to suburban locations. You can see how you like rabbits—and how well they like you—for a relatively small investment.

This is what this book is all about: We want to show you how to produce rabbits for family food or as a profit-making enterprise if you choose to do so. It is a complete guide to selection of stock, housing, breeding, feeding, slaughtering and above all, developing a market for the tasty nutritious meat.

There's a great opportunity for growth in the rabbit business. Progress requires enthusiastic breeders who will provide a consistent supply of rabbit meat. They must develop the market with a quality product, and one that keeps people coming back for more when they find out how good rabbit really is.

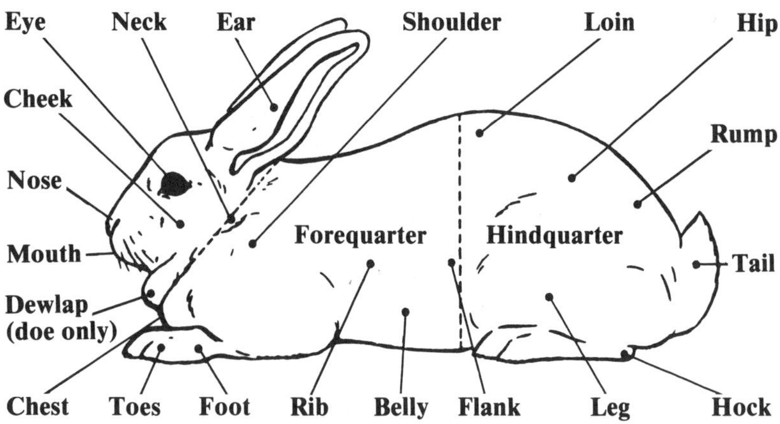

Parts of a Rabbit

A New Zealand doe with part of her litter

Chapter 1

Set your goal—food, profit or companionship

The first step in raising rabbits is to decide what you want to accomplish. People raise rabbits for many different reasons or a combination of reasons. The path you take determines the breed you select and the equipment you will need. You may very well take a different fork in the path after you learn the business. However, the place to start is with a clear-cut goal for the years immediately ahead.

Let's examine the possibilities:

Food for the family and friends: You start off with a buck and two or three does. They'll give you a rabbit a week or better for the family table.

Part-time commercial production: Here you are thinking about 50 does and five or six bucks. Your time investment will be 10 hours per week and you will have a sideline that produces both family meat and profit. You sell live rabbits or process the meat and market it.

Full-time commercial production: It requires around 500 does to make a living. Management of a herd of this size is a full-time job. A commercial processor is necessary to handle the production of a unit of this size.

Fanciers: Here your goal is to develop outstanding animals of a breed or breeds. You are breeding for perfection. Your income is built from the sale of breeding stock. Most members of the American Rabbit Breeders Association are in this end of the business. They do sell meat from the rabbits that do not quite measure up to top standards.

A 4-H Club project: Thousands of boys and girls raise rabbits as a 4-H Club project. They begin with the starter unit of a buck and two or three does and go on from there. In Ohio alone, there are over 5,000 4-H rabbit production members and the number is growing. These are the commercial rabbit producers of tomorrow.

Pets: Some people raise rabbits as pets. The kids love their bunnies and enjoy having them around. But when the kids fall in love with a rabbit, you have a permanent guest. They give bunnies cute names and make them a member of the family. There's no chance those rabbits will find their way into the family frying pan. Don't get attached to your rabbits...keep the same distance you would with pigs, chickens or cattle.

Where to start

We recommend that anyone not experienced in raising rabbits begin with our suggested starter unit of a buck and three does. That's big enough to see how well you like rabbits...and how well they get along under your care. This unit alone may produce more than 100 fryers per year. Regardless of your leanings...commercial production or the fancy breeds, it's wise to get experience before you invest big bucks in the business.

Once you have decided you want to get serious about rabbits, the first action is to decide whether or not zoning permits raising them. While rabbits fit into an urban setting very well, zoning may not permit a commercial rabbitry...or even a few pets for that matter. Check with your city or county planning commission. And it probably is a good idea to discuss it with the neighbors.

Income projections

The starter units we've been suggesting are not likely to be profitable when measured in dollars. The scale of production just isn't large enough to generate sufficient income. Your satisfaction comes from home-grown meat and a project that the whole family can enjoy. For profit, you need the 50-doe part-time unit we listed earlier. Following are some projections to show what it might produce for a family.

The 50-doe unit

In this unit, there are 50 does and six bucks. We have developed figures as a forecast of potential. They can be achieved with good management, but are not guaranteed. Cost figures vary according to feed, building and equipment price changes. The income represents a labor investment of about 10 hours per week. The following information can help you know what to expect from a rabbitry of this size.

Equipment Cost

Caging	$500.00
Feeders	110.00
Nest Boxes	125.00
Automatic watering system	150.00
	$885.00

Facts used in forecasting

1. First breeding time at 5 months of age.
2. 31-day gestation period. (29-32).
3. Wean litter at 8 weeks old, direct to market.
4. Doe is rebred 5 weeks after kindling.
5. Doe kindles again 1 week after weaning.
6. Kindling cycle averages 9 weeks—resulting in an average of 5 litters per year per doe.
7. Utilize 1 buck per every 10 to 20 does.

Feed consumed—facts used

Herd purchased at 8 weeks old.

From 2-3 months of age, the average body weight will be 5 pounds.

From 3-4 months of age, the average body weight will be 7 pounds.

From 4-5 months of age, the average body weight will be 9 pounds.

A growing rabbit will consume about 7% of its body weight in feed daily, decreasing down to an average of 4% by maturity.

2-3 months—5 lbs. @ 7% = .35 lbs. (6 oz.) per day x 30 days = 180 oz. or 11 lbs. per month.

3-4 months—7 lbs. @ 7% = .49 lbs. (8 oz.) per day x 30 days = 240 oz. or 15 lbs. per month.

5 months + – 9-10 lbs. @ 4% .... averaging 2.8 lbs per week.

2-3 months........11½ lbs.
3-4 months........15 lbs.
4-5 months........15 lbs.
41½ lbs. feed consumed from purchase to first breeding.

First Year Plan

Marketing dates

A calendar year is being used to make it easier to project returns. No sales until July 1.

Purchase herd on January 1 at 8 weeks of age (date of birth—November 1)
First breeding—April 1 thru April 30
Kindle—May 1 thru May 30
First marketing—July 1 thru July 30
Rebreed—Mid-June thru Mid-July
Kindle—Mid-August thru Mid-September
Second Marketing—Mid-October to Mid-November
Rebreed—October 1 thru November 1
Kindle—November 1 thru December 1
Third marketing—January 1 thru January 30

Gross income—50 working does

Averaging 8 young per litter x 3 litters per doe x 50 does = 1200 young.
1200 young, each averaging 4 lbs. = 4,800 lbs. marketable weight @ 60¢ per lb. = $2,880.
Manure sales (1 ton average per working doe per year) = 50 tons @ $10.00 = $500.

Feed consumption—50 working does

41½ lbs. x 56 = 2,324 lbs. or about 1¼ tons @ $240 = $300.
Herd from April 1 thru December will consume an average of 6,272 lbs. of feed (2.8 lbs. per week x 40 weeks = 112 lbs. x 56) or 3⅛ tons @ $180 = $759.60.
1200 young, each consuming about 6 lbs. of feed to reach market weight = 7200 lbs. of feed or 3.6 tons @ $240 = $864.

From purchase to April 1	$300.00
From April 1 thru December	759.60
Consumed by young from birth to market	864.00
Feed consumed for the year	$1,923.60

First year summary

50 working does	
Fryer sales	$2,880.00
Manure sales	500.00
Gross sales	3,380.00
Feed costs	– 1,923.60
Gross profit	$1,456.40

Second year and thereafter projected net

Gross income—50 working does

Averaging 8 young per litter x 5 litters per year x 50 does = 2,000 young.

2,000 fryers, each averaging 4 lbs. = 8,000 lbs. marketable weight @ 60¢ per lb. = $4,800.

Manure sales (1 ton average per working doe per year) 50 tons @ $10 = $500.

Feed Consumed—50 working does

Each young will consume approximately 6 lbs. of feed from birth to market. 2,000 young x 6 lbs. of feed = 12,000 lbs. feed.

Breeding herd of 56 will consume approximately 2.8 lbs. of feed each per week. 56 breeders x 2.8 lbs. feed = 156.8 per week x 52 weeks = 8,154 lbs. per year of consumed feed.

12,000 lbs. feed for young
8,154 lbs. feed for herd

20,154 lbs. @ 1.2¢ lb. = $2,418.

Second year summary

50 working does	
Fryer sales	$4,800.00
Manure sales	500.00
Gross sales	5,300.00
Feed cost	– 2,418.00
Gross Profit	$2,882.00

This figures out to be about $56 per doe.

Cost factors do not include replacement stock for your herd. Under normal circumstances, these costs will be offset by the sale of breeding and laboratory stock which command a higher price.

Not a get rich scheme

Please bear in mind, this is not a "get rich quick scheme." It takes hard, dedicated work before you can relax and enjoy the profits...If you don't have the "stickability" to stay in there and fight when the going gets tough...don't even consider becoming part of this industry.

This projection is designed to give you the highlights of the rabbit meat production business. The fine points and subsistence methods were not covered. The rabbit meat business is a serious business: you cannot earn a profit in any business without approaching it in this light.

How about a larger unit?

Economies of scale should make it possible to produce rabbit meat at less cost, as you go to a full-time production. However, equipment is more expensive and must be amortized. And remember, you must take a full-time salary out of the business. If you can make a gross profit of $56 per doe, as in our example, a 500-head herd should produce $28,000 per year gross profit. There are many overhead items, equipment amortization, taxes and other items to subtract from this before you come down to net income.

Other income opportunities

The sale of pelts can boost income a bit in the larger rabbitry. Even cotton tails and "lucky" rabbit's feet have made extra money for producers.

Laboratories buy large numbers of rabbits for their research needs. This is a good, but specialized market, that can be developed.

The best way to boost profit is to sell a processed rabbit at the front gate. Instead of the 60¢ per lb. you might receive from a live rabbit, you could expect to sell meat for $1.75 per lb. or more. The live rabbit at 60¢ brings in $2.40. The same rabbit will make two lbs. of dressed rabbit that returns $3.50. The supermarkets these days charge around $2.25 per lb. for a frozen rabbit. The more of this consumer dollar you can keep at home, the greater the return.

We have explained market opportunities in much greater detail in Chapter 7. Study that chapter as a guide to ways you can market your rabbits for meat, breeding stock for fur and novelties.

A long-eared French Lop on the judging table

Income from fancy rabbits

Profit from raising fancy rabbits for show is hard to measure. Not many breeders are in it primarily for the money. They get their satisfaction from developing better individual rabbits. They enjoy the competition at shows and the association with other breeders. The annual national show, for example, attracts more than 5,000 entries.

Winners and their offspring sell very well; sometimes for hundreds of dollars. In general, the range for breeders can be from $5 to $100. You can expect to pay $15 or so for your commercial breeding stock from one of these foundation stock producers.

Like any other business...the good ones do very well. The breeders with less success fade away. Right or wrong, almost any fancy breeder you talk with will tell you his business is better than raising meat animals.

We've given you the basic information to help you decide the course of your entry into the rabbit business. Again...start small and grow when you know you will enjoy raising rabbits and can make a financial success of your venture.

Lightweight rabbits (l to r): Dutch, Polish, Havana, Silver Marten, Harlequin, Rex

Chapter 2

Select the breed that fits your plans

Rabbits come in many sizes, colors and capabilities. The first step in developing a sound rabbit business is to choose quality individuals of the breed that best fits your purpose.

In 1975, the American Rabbit Breeder's Association (ARBA) recognized 38 different breeds. Many of these are color varieties of the same basic rabbit. Although the number may sound confusing, the great majority are for fanciers. If you are in the rabbit business for profit from meat animals, the choice is relatively limited.

There are three general sizes of rabbits; small, medium and large. The small rabbits are popular on the show circuit and make good pets. The mature rabbit is too small for the meat market. The medium-sized rabbits are best for meat. They reach four pounds in eight weeks or less. The large rabbits do not grow quickly enough to be profitable meat animals. And the carcass from a 15-pound rabbit is not in as much demand as that from a medium rabbit.

The best meat breed

For the meat business, the New Zealand White probably is your best choice. It is a very popular rabbit and that means stock is readily available. The Californian is another excellent meat rabbit.

These breeds develop a quality carcass and they produce large litters consistently. White-furred rabbits are usually preferable since the rabbit fur market often brings a higher return for white pelts that can be dyed. The laboratory trade also prefers a white rabbit.

If you decide on a specialty-type rabbit to sell to other breeders or as pets, select a breed that you can successfully raise in your rabbitry. Choose one you think will sell easily.

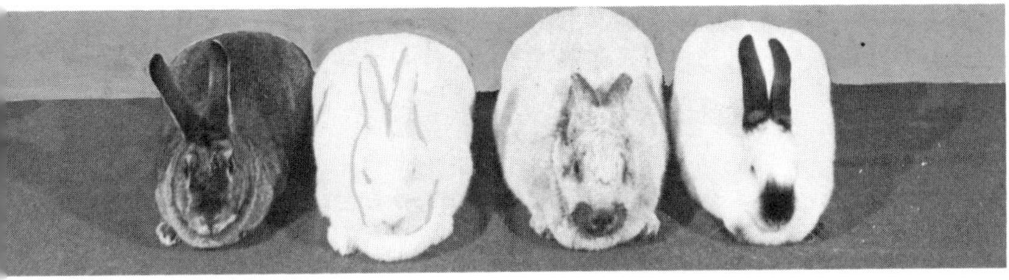

Heavyweight rabbits (l to r): Satin, New Zealand, Champagne D'Argent, Californian

For the pet market, the preferred breeds are those that are small, come in various colors and patterns. Select those that are easy to raise, can be weaned early, and do not tend to become mean or flighty when handled. Pets must be capable of adjusting to unusual conditions. Many animals sold as pets are marketed soon after weaning or sometimes even before they are fully weaned.

If you plan to sell rabbits to breeders or fanciers, the rabbits must fit the official standard description for that breed as closely as possible. Fanciers usually decide on a breed because they admire specific qualities of that breed. Breeders often require animals that are registered or that can be registered.

Four types of fur

The rabbit breeds are grouped into four fur types—normal, satin, rex, and angora. Most breeds of rabbits have normal-type fur.

Normal fur has a dense undercoat protected by longer guard hairs. Satin fur is finer than normal fur, is composed of transparent hair, and has a sheen or luster. Rex fur is very dense and the guard hairs are approximately the same length as the undercoat, giving a plush feeling. Angora fur contains a long, wool-type fiber that is harvested and used in the manufacture of clothing.

The chart on the next two pages shows the standard breeds and varieties of rabbits as rated by the ARBA. You'll note that only two are rated "very good" for meat production. They are the Californian and the New Zealand.

Close-up of the major rabbit breeds

The rabbits pictured on the following pages are among the most popular in the United States today. Their history, characteristics and place in the rabbit market are described.

Standard Rabbit Breeds and Varieties

Breed	Variety
Alaska	Black
American	Blue, White
Angora, English	Black, Blue, Fawn, White
Angora, French	Black, Blue, Fawn, White
Belgian Hare	
Beveren	Black, Blue, White
Californian	
Cinnamon	
Champagne D'Argent	
Creme D'Argent	
Checkered Giant	Black, Blue
Chinchilla, American	
Chinchilla, Giant	
Chinchilla, Standard	
Dutch	Black, Blue, Chocolate, Gray, Steel Gray, Tortoise
English Spot	Black, Blue, Chocolate, Gold, Gray, Lilac, Tortoise
Flemish Giant	Black, Blue, Fawn, Light Gray, Sandy, Steel Gray, White
Florida White	
Harlequin	Harlequin, Magpie
Havana	Chocolate, Blue
Himalayan	
Lilac	
Lop, English	Broken, Solid
Lop, French	Broken, Solid
Netherland Dwarf	Agouti, Patterned, Selfs, Shaded, Tan, Any Other Variety
New Zealand	Black, Red, White
Palomino	Golden, Lynx
Polish	Black, Chocolate, Blue-eyed White, Ruby-eyed White
Rex	Black, Blue, California, Castor, Chinchilla, Chocolate, Lilac, Lynx, Opal, Red, Sable, Seal, White
Rhinelander	
Sable	
Sable, Siamese	
Satin	Black, Blue, Californian, Chinchilla, Chocolate, Copper, Red, Siamese, White
Silver	Brown, Fawn, Gray
Silver Fox	Black, Blue
Silver Marten	Black, Blue, Chocolate, Sable
Tan	Black, Blue, Chocolate, Lilac
Vienna, Blue	

Pounds Ideal Mature Weight		Utility Value		Popularity for Show[1]
buck	doe	meat	fur	
7½	7½	fair	fair	poor
9	10	good	good	poor
6	7	good	good	fair
8	8	fair	fair	fair
8	8	fair	fair	poor
9	10	good	good	fair
9	9	very good	very good	very good
9	10	good	fair	fair
10	10½	good	good	good
9	10	good	good	fair
11 +	12 +	good	fair	good
10	11	good	good	fair
13 to 14	14 to 15	good	good	fair
6½	7	good	good	fair
4½	4½	small	fair	very good
6	7	small	fair	fair
14 +	15 +	good	good	fair
5	5	small	good	poor
7	7	fair	good	poor
6	6	small	good	fair
3½	3½	small	fair	poor
6 to 7	6½ to 7½	good	fair	fair
10	11	good	fair	fair
10	11	good	fair	good
2	2	small	fair	good
10	11	very good	very good	very good
9	10	good	good	fair
2½	2½	small	fair	fair
8	9	good	good	good
8½	9	fair	fair	new
8	9	good	fair	poor
5 to 7	5 to 7	small	fair	poor
9½	10	good	good	very good
6	6	small	fair	poor
10½	10½	good	good	fair
7½	8½	good	good	good
4 to 5	4½ to 5½	small	good	fair
9 to 10	10 to 11	good	good	poor

[1] Based on number of exhibitors and number shown in shows during several years.

New Zealand

The New Zealand Red, the New Zealand White, and the New Zealand Black Rabbits are strictly American creations; their names have nothing to do with their origin. Their true ancestors are the Belgian Hare and the Golden Fawn, both popular breeds before the New Zealands made their appearance. First there were the Belgian Hares, then the Golden Fawns. After that, the New Zealand Reds and, eventually, the White made their appearance. American breeders who created and perfected the New Zealand breed made a wonderful contribution; these rabbits are now finding their way into European countries, just as other breeds formerly found their way from Europe into America. The New Zealand is considered a commercial breed. Its adult weights are 9-11 pounds for bucks and 10-12 pounds for does.

Californian

The Californian is a medium-sized, meat-type rabbit, originally developed in the Golden State. It usually is white with black or gray ears and other extremities. This rabbit is considered to have the ideal carcass type. It is extremely broad with a high yield of meat. The Californian is prolific and produces large litters. Adult weights are 8-10 pounds for bucks and 8½-10 pounds for does. Since it is not as popular as the New Zealand, breeding stock may be more difficult to locate.

Champagne D'Argent

Champagne D'Argents, also known as French Silvers, are probably one of the oldest known breeds. They have been raised in France for more than 100 years. These rabbits were originally raised in great quantities, and they doubtlessly originated in the Champagne district of France, from which the breed derives its name. The fur is useful in its natural state and is one of the leading furs used in garment manufacture throughout Europe. Mature bucks weigh 9-11 pounds, and mature does weigh 9½-12 pounds. Ideal weights for bucks are 10 pounds and 10½ pounds for does. Their under color is dark slate blue, and their surface color is as near the color of skimmed milk or light aquamarine (a washed-out blue) as possible. The body fur has a liberal sprinkling of longer black guard hairs. Born black, rabbits of this breed gradually turn silver as they mature.

The French Lop

There are two species of lop eared rabbits, the English and the French. It is said they first appeared in Algiers, North Africa, and then spread to France, Belgium, and the Netherlands. England adopted them as an ideal exhibition rabbit because of their extraordinary ears which measured 24 inches from tip to tip. Ear lengths of 28½ inches have been attained. This is one of the rarer breeds, especially in America. Ears are much shorter than are those of the English variety and not as wide. This is more of a utility rabbit in the 10-11 pound class.

Himalayan

This is one of the oldest established breeds, with wider distribution throughout the world than any other rabbit, being short, soft, and pure white except for the ears, nose, feet, and tail which should be a rich, velvety black.

Angora

The Angora produces wool of the finest quality rather than fur. Its utility value has long been recognized by the English and French people. There is some difference in the English and French versions of this rabbit—the length and texture of the wool distinguishes one from the other. For many years, the French in the Visges, Savoy, Mayence, Normandy, and Jura districts spun the wool and knitted it into the finished garments in their own homes. In recent years, both in America and abroad, large commercial mills have taken over the manufacture of these garments on a large scale. These garments are extremely lightweight, soft, and beautiful.

The French Angoras come in the same colors as the English, but they are larger, having a medium length body. The ideal weight is 8 pounds for the French Angora. If well-covered with good quality and length of wool, the French will give the appearance of a large, fluffy, oval shape. White English in same position and condition would have the appearance of a fluffy ball. With French, as with English, uniform length, density, and texture of the wool is of vital importance. It is important that condition, length, density, and texture of wool be given first and most consideration in breeding and exhibition.

Chinchillas

Chinchillas exist in three breeds, based on size. They are beautiful animals, having fur that resembles the South American chinchilla, from which they take their name. The fur should be even in color over the entire exposed surface of the body. The color is a bluish gray in total, being made up of dark blue at the skin with intermediate layers of white and an outer layer of dark bluish gray. Above this is a layer of near-white tipped with black. This makes for a ringed appearance when the hair is parted and an overall stippled effect when the hair lies normally. Chinchillas are normally bred as commercial meat rabbits, and the carcass of the better specimens is of good meat confirmation (shape, size, and proportion). The fur is marketable. The Standard variety, because of its small size, is usually considered a fancy breed. The Giants appear bolder and more powerful and weigh 14-15 pounds.

Dutch

Dutch rabbits have very distinctive markings. The standard colors are black, blue, chocolate, tortoise, steel gray and gray. All colors have a white shoulder-foreleg area, white back feet and distinctive coloration of eyepatch and ear. Dutch are most attractive rabbits. Their 4½ pound size makes them ideal hobby and show animals. They also are in demand in the laboratory market.

The Flemish Giant

The Flemish Giant will always be a popular breed. Colors include steel gray, light gray, sandy gray, blue, black, white and fawn. These rabbits are sponsored by the oldest and one of the strongest specialty clubs in the nation. As the name "Giants" implies, they should be as large as possible, have a powerful, massive build, and must be properly proportioned and well-balanced throughout. Minimum weights are 12 pounds for bucks and 13 pounds for does and as much more as possible, always keeping in mind a well-balanced, well-proportioned animal throughout. Many reach 20 pounds.

Netherland Dwarf

This breed is fairly new in this country, imported primarily from Holland and England. It's a strictly fancy breed, with a top weight of 2½ pounds. These tiny rabbits have gained considerable popularity in a short time. Being exceptionally small, they attract interest at shows. They have a compact body, globular head and small, upright ears. There are five categories of colors. As a small rabbit, the Netherland is very economical to feed.

Silver Marten

This rabbit, which had the Chinchilla as its forerunner, is not difficult to produce. When these "sports" (genetic mutations) from the Chinchilla rabbit are bred together, they usually breed true to type and color, seldom reverting to Chinchillas. Silver Martens come in blue, black, chocolate, and sable. Ideal weights for all colors are 7½ pounds for bucks and 8½ pounds for does. These weights, together with the compact body and meaty back and shoulders, make a good commercial rabbit.

Belgian Hare

This rabbit has slim, graceful body lines. Its standup position, together with an alert expression, make this one of the most spectacular of all rabbit breeds. No other breed possesses as much fascination for the out-and-out fancier as does the Belgian Hare.

This is the race horse of the rabbit family. It has a long, fine body, nicely arched with a continuous sweep from shoulders to tail.

Satin

The Satin rabbit is an American rabbit characterized by its very dense fur with a high degree of sheen. The fur also is of very fine texture. It is a medium-sized rabbit that produces good quality meat. Bucks average 9½ pounds and does 10 pounds. It is produced in a variety of colors and the attractive fur makes it a good show animal.

Rex

The Rex rabbit is a medium-weight rabbit very much in demand for its fur. The fur is relatively short, about ½-inch in length and should be of uniform texture over the entire body. It is free of guard hairs and has the appearance and feel of velvet. Rex fur makes beautiful coats and other garments. These rabbits come in various colors: black, blue, castor, Chinchilla, chocolate, lilac, lynx, opal, red sable, and white. One of the problems is obtaining enough furs of uniform color to make a garment. After the fur is removed, the Rex makes a good meat animal. Bucks weigh around 8 pounds; does 9 pounds.

Before you get into the rabbit business, talk with other breeders. Get their ideas on the best breeds for your situation. Go to a rabbit show. See the top animals in each breed. That's the way to get acquainted.

For information about any breed, write to American Rabbit Breeders Association, Inc., 1925 Main St., Bloomington, IL 61701. There are specialty clubs for all of the major rabbit breeds. The ARBA can help put you in touch with any one of the 37 specialty clubs it has on its roster.

Remember, the right breed for you is the key to success. Study them carefully before making your choice.

A pair of Broken Rex rabbits

Chapter 3

How to get started in rabbits

The first rabbits you buy will become the firm foundation on which you build your future herd. It may take generations to overcome the heritage of a poor choice made at the beginning. That's why it is so important to make the best selection available.

Once you have decided upon the breed of rabbit that fits your objective, the next is to select the stock to start. Hopefully, you have chosen a breed that is reasonably popular in your area. This gives you an opportunity to visit several breeders to see their animals and to inspect their records.

My personal choice would be to buy from a breeder within an hour's drive of home. There's less stress on your new stock in transit. And that breeder is handy to answer questions and to back up his stock should problems arise.

Pick up any rabbit magazine and you will find classifieds offering rabbits. Many mail-order rabbits are of excellent quality. And they do make the trip in good health. But, on the other hand, distance does enhance the opportunity for problems.

The common meat breeds, such as the New Zealand or Californian, usually are available within a short distance. There are many breeders of these rabbits. If you select a fancy breed, you may have to go to a distant source. There are only a few breeders of some of these rabbits, scattered throughout the United States.

Ship by air

These days, air freight or an airlines package express probably are the best ways to ship. Rabbits can be delivered to your closest airport in one day or overnight. Costs are a minimum of $25. Even $50 isn't excessive for three or four rabbits of the special breeding you want.

Another source is the broker. He sells you rabbits he has purchased from another source. He usually relies on high-pressure advertising to pull in customers. Remember, this source is a salesman who probably doesn't know much about the rabbits he is selling you . . . and probably doesn't care very much about quality. We suggest that you avoid the broker in your search for foundation stock.

A rabbit show is a good place to locate breeders—but not to buy. You'll find a selection of rabbits of many breeds. Compare your choices with others. You can talk to the owners and actually see for yourself both the rabbits and the man who sells them.

The next step is to actually visit the rabbitry and pick out the foundation herd. Tell the grower what you want to do in the rabbit business. Ask him for help in selecting the rabbits he feels will do the job for you.

Look around the operation. Are the pens clean? Do the rabbits look healthy? Are they in good flesh? Do they come to the front of the hutch when the owner approaches? Watch the owner. If he appears to love his animals, if he handles them gently and perhaps talks to them, you will usually conclude that you have come to the right place to buy your rabbits.

Records are just as important in selecting rabbits as they are for livestock and poultry. Today's livestock producer insists on knowing the performance of the percentage before he selects foundation stock. You should be equally cautious on your selection of does and bucks to start your rabbitry.

Information you'll need

Get as much background information on each rabbit and its ancestors as possible. Read a breeder's records to find out:

- If any of that rabbit's family has been registered.
- The production for both the bucks and does.
- Production of weaned rabbits.
- The rate of weight gain.
- The number of animals sold as breeders.
- Both the desirable and undesirable characteristics of the stock.

Be sure a doe breeds easily, conceives readily, produces adequate milk, makes a good home for her young and raises a uniform litter. You'll want her to continue in production and produce offspring as good as, or better than, herself. Be sure that this stock will produce the quality of rabbit you want. Invest in a rabbit from a mother that has raised at least 30 fryers (young rabbits) a year. If you are not buying pedigreed stock, then be sure your rabbit is at least typical of its breed, especially in coloring and size.

A New Zealand buck

The buck is very important

Give even more attention to the selection of a buck. Remember that half of the characteristics of your young bunnies come from the buck. If you have one buck and 10 does, that buck has 10 times more impact on the herd than any one of the does. Ask questions about the bucks heritage. Make sure he comes from a line of big litters of fast growing young.

If you plan to raise meat-type stock, look at the body conformation of the animal. Look for an animal that has a medium, blocky, deep-bodied appearance with broad, smooth hind-quarters. Watch for an alert appearance, reasonable fur quality and well-padded feet.

Regardless of breed, only select animals that appear to be outstanding examples of the type you want. Select stock that is in healthy condition. Do all you can to ensure the least amount of stress possible on the animals when moving them to your rabbitry.

What age for starters?

Some breeders have started with mature animals. You can get a head start with a proven breeder. However, you always must ask yourself, "why is this doe for sale?" A breeder will want to retain his best animals for his own herd. The mature doe may not be your best buy.

When we speak of a mature doe we mean one that has been bred and will shortly have a litter. Perhaps she already has had one or more litters. A bred doe costs more than an unbred one and is hard to find. When bred does are not available, you may wish to purchase animals that are 8 weeks old since these animals cost less and are easier to find. In addition, the rabbits have time to adjust to their new home and you have time to learn about the new animals before breeding. Remember, our recommended starter set is three does and a buck.

Do you want "papers?"

You can select rabbits with a pedigree and registration papers. This is the best way to insure that the rabbits you select have the breeding and background you want.

What is a pedigree?

The pedigree is most useful as a genealogical record. Since the breeding of good rabbits is quite dependent upon the mating of genes from the two parent animals in the progeny, a breeder must know what genes he is uniting. The only means to know this presently, is to study the pedigree. Many pedigree forms are in use today, and some

Ear No. ______ Breed ______ Sex ______ Birth ______
Selling Price ______ Sire Name ______ Dam Name ______
Date ______ Name ______ Address ______ City ______ State ______ Zip ______

ORIGINAL

A.R.B.A. Rabbit Pedigree

EAR No. R2-D

SOLD TO
NAME Steven J. McHenry
ADDRESS 143 School Street
CITY Des Moines STATE IA
DATE SOLD 11-2-80
BREED Am. Chincilla
SEX Buck
BORN 9-15-80

NAME Rufus II
Registry No. 9906 D
Color Brown Weight 4
Grand Champ. No.
Winnings:

SIRE Reids' Af
Registry No. 9844C
Color Brown Weight 9-1/2
Grand Champ. No.
Winnings:

Grand SIRE Kellers' 25-G5
Reg. No. 9128B
Color Brown Weight 9
Grand Champ. No.
Winnings:

G. G. SIRE Krugs' 8C3
Reg. No. 3631B
Color Brown Weight 9
Grand Champ. No.

G. G. DAM Krugs' EG8
Reg. No. 5150B
Color Brown Weight 10
Grand Champ. No.

Grand DAM Reids' D-1
Reg. No. 6866B
Color Brown Weight 10
Grand Champ. No.
Winnings:

G. G. SIRE Reids' NR2
Reg. No. 6847B
Color Brown Weight 9-1/2
Grand Champ. No.

G. G. DAM Reids' R2-C
Reg. No. 6855B
Color Brown Weight 10
Grand Champ. No.

DAM Reids' A5
Registry No. 9854C
Color Brown Weight 10-1/2
Grand Champ. No.
Winnings:

Grand SIRE Kellers' 25-G5
Reg. No. 9128B
Color Brown Weight 9
Grand Champ. No.
Winnings.

G. G. SIRE Krugs' 8C3
Reg. No. 3631B
Color Brown Weight 9
Grand Champ. No.

G. G. DAM Krugs' EG8
Reg. No. 5150B
Color Brown Weight 10
Grand Champ. No.

Grand DAM Reids' D-1
Reg. No. 6866B
Color Brown Weight 10
Grand Champ. No.
Winnings:

G. G. SIRE Reids' NR2
Reg. No. 6847B
Color Brown Weight 9-1/2
Grand Champ. No.

G. G. DAM Reids' R2-C
Reg. No. 6855B
Color Brown Weight 10
Grand Champ. No.

I hereby certify that this Pedigree is correct to the best of my knowledge and belief.

NAME
ADDRESS
CITY STATE

THE AMERICAN RABBIT BREEDERS ASSN. Inc. MEMBER

are of no value in determining the mating of animals.

A pedigree should give a brief description of every rabbit for three generations, and many breeders can trace back several years for as many as 25 generations of such descriptions. On the other hand, some forms list only a name or a few numbers.

What is registration?

Registration of a rabbit can mean a great deal. A rabbit must first have a pedigree, then must conform to rigid standards before being registered. This can only be done by contacting an ARBA-licensed registrar who will thoroughly examine the animal before making application for registration. Cost is $3 per rabbit.

Registration denotes quality in a rabbit. When all animals on a pedigree are registered, you can be sure all ancestors were at least minimum weight and no poor specimens contributed to the background of that particular animal. Marked breeds are especially difficult to keep up to standards; in addition to weight, health, and lack of deformities, the animals must all have specific markings in just the right places.

The primary purpose in registration, then, is to insure good breeding stock. Breeders who do not register their herds often keep poor does in the breeding program and will not register other rabbits since they do

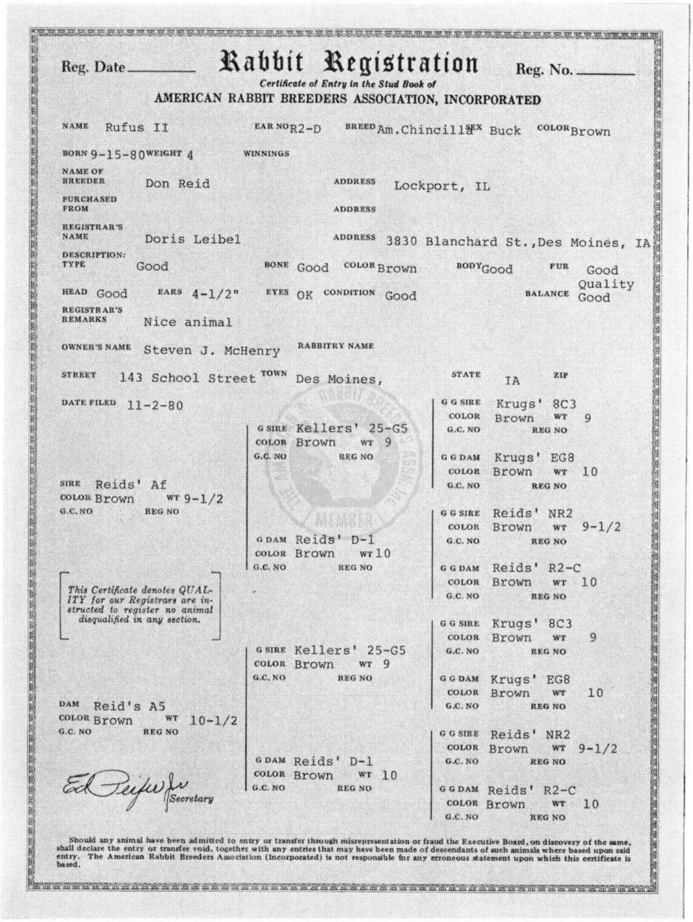

Rabbit Registration

Reg. Date ______ Reg. No. ______

Certificate of Entry in the Stud Book of

AMERICAN RABBIT BREEDERS ASSOCIATION, INCORPORATED

NAME Rufus II EAR NO R2-D BREED Am.Chincilla SEX Buck COLOR Brown

BORN 9-15-80 WEIGHT 4 WINNINGS

NAME OF BREEDER Don Reid ADDRESS Lockport, IL

PURCHASED FROM ADDRESS

REGISTRAR'S NAME Doris Leibel ADDRESS 3830 Blanchard St.,Des Moines, IA

DESCRIPTION: TYPE Good BONE Good COLOR Brown BODY Good FUR Good Quality

HEAD Good EARS 4-1/2" EYES OK CONDITION Good BALANCE Good

REGISTRAR'S REMARKS Nice animal

OWNER'S NAME Steven J. McHenry RABBITRY NAME

STREET 143 School Street TOWN Des Moines, STATE IA ZIP

DATE FILED 11-2-80

SIRE Reids' Af COLOR Brown WT 9-1/2 G.C. NO REG NO

G SIRE Kellers' 25-G5 COLOR Brown WT 9 G.C. NO REG NO

G G SIRE Krugs' 8C3 COLOR Brown WT 9 G.C. NO REG NO

G G DAM Krugs' EG8 COLOR Brown WT 10 G.C. NO REG NO

G DAM Reids' D-1 COLOR Brown WT 10 G.C. NO REG NO

G G SIRE Reids' NR2 COLOR Brown WT 9-1/2 G.C. NO REG NO

G G DAM Reids' R2-C COLOR Brown WT 10 G.C. NO REG NO

This Certificate denotes QUALITY for our Registrars are instructed to register no animal disqualified in any section.

DAM Reid's A5 COLOR Brown WT 10-1/2 G.C. NO REG NO

G SIRE Kellers' 25-G5 COLOR Brown WT 9 G.C. NO REG NO

G G SIRE Krugs' 8C3 COLOR Brown WT 9 G.C. NO REG NO

G G DAM Krugs' EG8 COLOR Brown WT 10 G.C. NO REG NO

G DAM Reids' D-1 COLOR Brown WT 10 G.C. NO REG NO

G G SIRE Reids' NR2 COLOR Brown WT 9-1/2 G.C. NO REG NO

G G DAM Reids' R2-C COLOR Brown WT 10 G.C. NO REG NO

Secretary

Should any animal have been admitted to entry or transfer through misrepresentation or fraud the Executive Board, on discovery of the same, shall declare the entry or transfer void, together with any entries that may have been made of descendants of such animals where based upon said entry. The American Rabbit Breeders Association (Incorporated) is not responsible for any erroneous statement upon which this certificate is based.

not want to be forced to keep only top quality breeding does. Others claim they cannot afford the extra expense. These cannot be excuses. Pedigrees and registrations are both important.

A.R.B.A. merit seal system. The national office of the American Rabbit Breeders Association makes up the certificate of registration, checks the registration numbers which appear on the pedigree, and affixes the merit seal. If both sire and dam are registered, a red seal is affixed.

If the sire and dam and great sire and dam are all registered, the red and white seal is affixed. If all ancestors showing on the pedigree are registered, a red, white and blue seal is affixed. This system is known as the merit system and shows how many generations had been pure bred and free from disease and defects.

Red Seal

Red & White Seal

Red, White & Blue Seal

Transfer of ownership. On the back of each certificate of registration issued by the association are three spaces reserved for transfer of ownership. The owner, when he sells the rabbit, signs his name on the top line and places the new owner's name and address below. This is then forwarded by either the new owner or the old owner to the national office with the fee of 25¢. This information is then transferred to the original registration application and is stamped, officially transferred, and the certificate in this form is returned to the new owner. It indicates he or she is now the owner of this registered rabbit.

The registration system is your safeguard in obtaining good rabbits that are healthy and free from serious defects. It is your duty to use it to the best advantage because, generally, an extra dollar or two spent, is many dollars saved.

Registration for meat animals?

If your goal is that of fancier, with emphasis on show and breeding stock sales, then registration is almost obligatory. If you are raising meat animals, then it may be better to go to a commercial breeder of grade animals.

The purebred with a showring blue ribbon is there because she has superior confirmation and appearance. She has not been judged on the basis of her ability to produce and mother big litters of fast-growing rabbits. A hard-working, milked-down mother won't be a show winner.

The beef and hog producers have gone through the same situation. The big winners in the showring, often have flopped when their carcasses were judged for meatiness on the hook. Now breeding stock is performance tested and carcasses are checked regularly.

The alternative for the new rabbit producer is to go to a commercial producer who has been selecting rabbits primarily on their ability to produce meat. If he has the records that show high performance, these are the rabbits that may do best for you.

Registered purebreds command a premium price when you select them for breeding stock. It costs money to maintain the paperwork and they should carry a higher price. And when you get them home, it will cost you money to continue the registrations.

Is this necessary for a rabbitry that concentrates on meat production? The difference may be between $10 for a grade rabbit and $20 for one with the papers. It's a basic decision you'll have to make, based on your own situation. Our suggestion is to keep up with the paperwork until it seems unnecessary.

A check list for rabbit selection

1. Buy from reliable breeders who keep records on litters and weights of stock.
2. Select from families producing big litters.
3. Select from families that are heavy milk producers.
4. Select from families that grow fast.
5. Select from does that wean four large litters a year.
6. Select from families showing resistance to disease.
7. Select from families with good type.
8. Select from families that live long.
9. Select stock free from undesirable characteristics (woolies, etc.).

What to look for in a rabbit

You can spot potential problems by taking a careful look at breeding stock before you buy. If a rabbit has weak characteristics, those traits may be transmitted to the young. Try to eliminate potential problems at the beginning.

Your rabbits should be free from defects or diseases. Do not buy a rabbit with a wry tail, rupture, buck teeth, flop ears, crooked legs, or bad eyes. These defects may be inherited; their young might have them too. Such conditions as ear canker, snuffles, sore hock, scours (dirty rears), vent disease, bloat, and others may be due to harmful bacteria (germs). If these diseases are brought to your place, they may spread to other rabbits. Remember, meat rabbits should be well developed in the important parts: the hind legs, hips, and loin.

The chart on the following page shows what to look for when you make a visual inspection.

Points to check when selecting breeders

The ideal rabbit

Well balanced rabbit throughout, good head, well carried ears, good bone, good type.

Good depth or rise to hips, very nice and smooth.

Avoid these problems

Belled ears, pear shaped head, double dewlap, back too flat and straight, sloped rump, cow hocked.

Broad hips but a little flat and not enough rise.

Narrow head, open carried ears, dip in back, cut off hips, pot bellied, side carried dewlap.

Shows protruding hip bones, or rough hips.

Correct tail carriage.

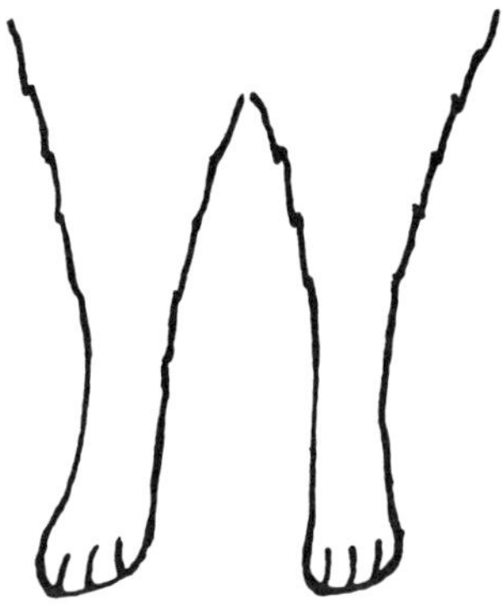

Strong straight limbs.

Screw tail.

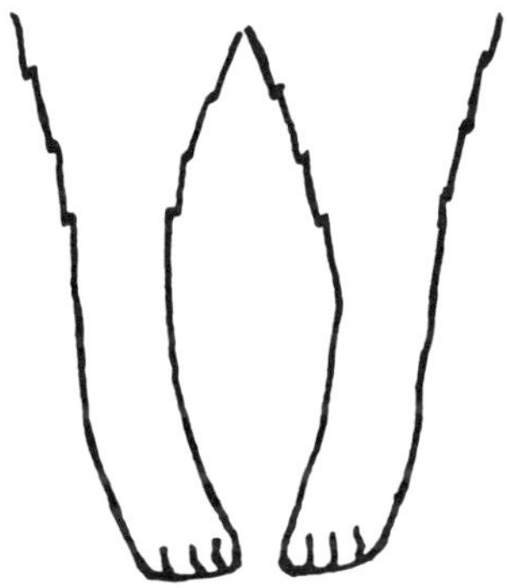

Bowed limbs, inward.

Side carried tail.

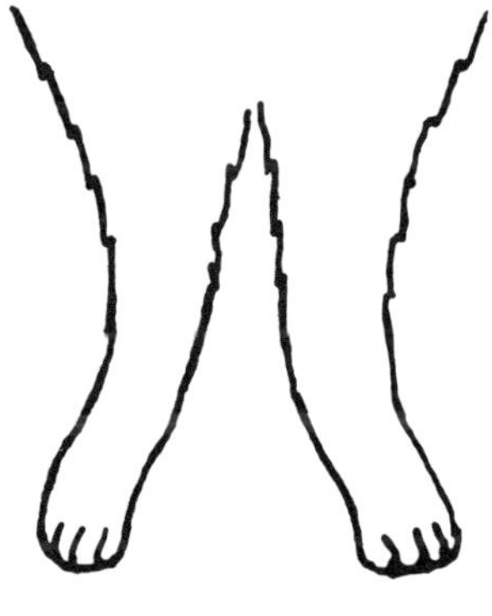

Bowed limbs, outward.

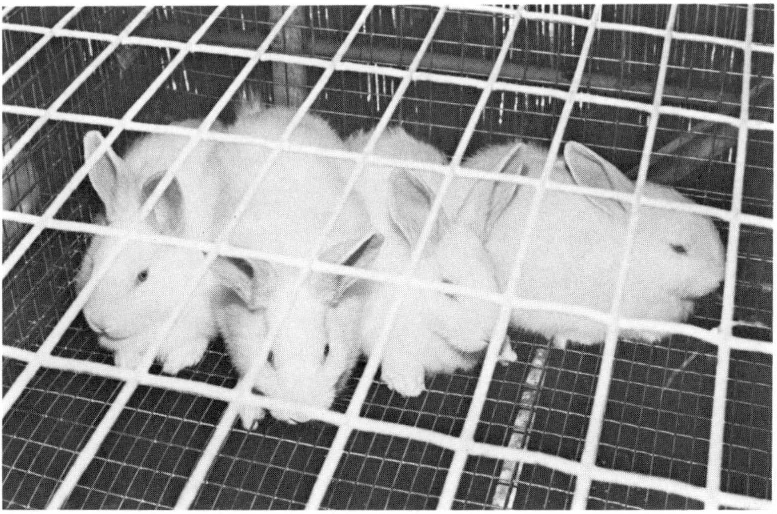

Today, almost all rabbits are raised in wire hutches.

Chapter 4

Housing and equipment you'll need

People who have been in the rabbit business for awhile, agree that one of the mistakes a beginner often makes is not being ready when the rabbits arrive. Being ready means adequate housing, hutches, feeders and waterers all ready for "guests".

It is a big shock for even a placid, domestic rabbit to move from one location to another. There's enough of the wild left in the rabbit to make it anxious and upset at a major move. Young rabbits often are just weaned and are getting used to new rations. That's why it is so important to have the rabbit's new home ready at the end of the move.

The rabbit and the environment

Rabbits do very well in adjusting to widely varying climatic conditions. The domestic rabbit, like its wild counterpart, can stand cold very well. However, it does not do so well in a drafty building. Dampness is almost sure to lead to a variety of health problems.

Heat is more of a problem for the rabbit than cold. Rabbits maintain their normal body temperature in hot months by losing heat through breathing. Their rate of breathing steps up as it gets hotter.

They also lose heat through the less densely furred areas of the body. Ears act as a "ventilator".

Since rabbit production is centered in Arkansas, Texas and California, the hot months frequently are a problem. Temperatures above 85° and high humidity often cause heat prostration and death losses. In a recent summer, when temperatures were above 100° for weeks on end, losses were extremely high. Breeding cycles were severely disrupted.

Another big problem in hot climates is reproduction. Bucks often refuse to breed in hot weather. Many never recover and are not effective breeders even when the weather cools off.

The bottom line is that rabbits raised in the hotter climates are going to need some "air conditioning" in their home. It may be fans, water spray on the roof or a simple evaporative cooler. Aluminum paint on the roof helps, too. There also will be days in the Midwest or East when rabbits will need some help.

In areas where strong winds and cold weather prevail, hutches need protection inside a building. Preferably, that building opens to the south or east. During the worst storms of winter, it should be possible to close these buildings. Many large producers heat their buildings to keep winter temperatures around 50°.

This is a typical converted broiler house in the south.

Adapt a building

More than 90% of all rabbits raised today are in a building that was originally built for some other purpose. In the Southern Plains, where commercial rabbit production is centered, broiler houses frequently are converted. Typically, two double rows of hutches run down the length of a long building.

In other areas smaller brooder houses or hen houses, left-over from the days when every family kept a small flock, have been converted for rabbits. The rabbit makes a good second-wave tenant. His needs are similar to chickens in many ways. And it is hard for the rabbit to pay for fancy surroundings.

The type of building you need for housing the hutches will be determined by local building regulations, climatic conditions, and the amount of money you can invest. In planning your building and its equipment, emphasize comfort of the rabbits and convenience for the caretaker. The building should have a simple design, protect the rabbits from winds, rain, and bright sun, and provide light and fresh air. Where mild climates prevail, hutches may be placed in the open but should have individual roofs and protection from the weather.

Sunlight helps maintain a sanitary condition in the rabbitry but whether it actually helps the rabbits themselves, has not been determined. Rabbits apparently enjoy being in the sun where temperatures are low or moderate but it is not necessary that they receive direct sunlight. In fact, exposure to direct hot sun may have a serious effect on the health and performance of your rabbits.

In mild climates, hutches may be placed in the shade of trees or buildings or man-made shade. This cuts off the hottest sunlight and helps keep rabbits from overheating.

Rabbits need protection

As we mentioned in the introduction, rabbits are a vital part of the food chain for many wild (and domesticated) animals. It is only natural that they seek out rabbits for food. Rabbits do best when they are not disturbed. Keep them in a quiet place, protected from dogs, cats, weasels and rats. Hutches must be constructed to give the necessary protection. Rats and weasels can squeeze through a surprisingly small opening. Therefore, mesh 1 x 2 inches, makes a good covering for the sides of a hutch. Use tough 16-gage wire on the floor.

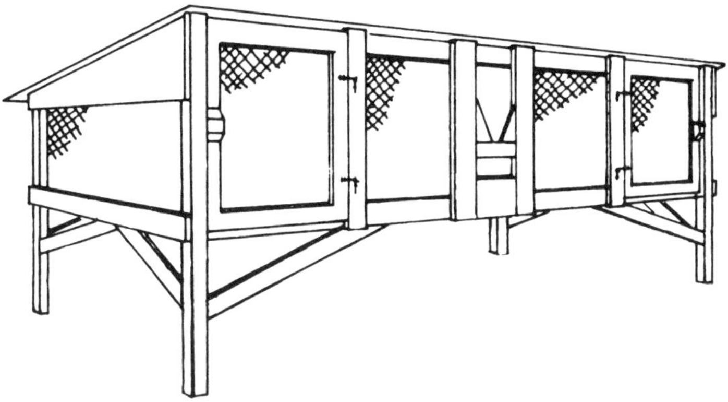

How to start

In this chapter, we will talk about housing and equipment for the new rabbit producers who want to start conservatively with the four rabbit unit we suggest. This keeps the cost down while you determine how far you really want to go in the rabbit business.

Later, we will talk about buildings and equipment for the commercial producer in the 50-doe part-time operation, up to the 500-doe plus full-time operation.

Basically, there are two ways to get started. One is with an outdoor self-contained hutch. These usually have wooden frames and wire cages. Two to four rabbits are kept in each of the hutches.

The units are portable. They can be moved around in the shade to keep the rabbits cool in the summer. In colder months, they can be moved inside for protection of the rabbits from snow and wind.

Another way to start is with steel wire hutches suspended from the joists or ceiling. Or they may be placed upon a framework inside an existing building. If you have average mechanical skills, you can make these cages with wire purchased from your local hardware store. You may find it only moderately more expensive to buy the units ready-built.

Avoid all-wood cages and hutches. These become urine soaked and tend to smell. They are difficult to sanitize between litters. And perhaps an even bigger problem is that rabbits tend to gnaw on the wood. From the standpoint of easy management, sanitation, good health and long life, stay away from the wooden hutches. Wood framework for a hutch is satisfactory...but keep the rabbits inside wire. Be wary of used cages...you may be buying the former owner's health problems.

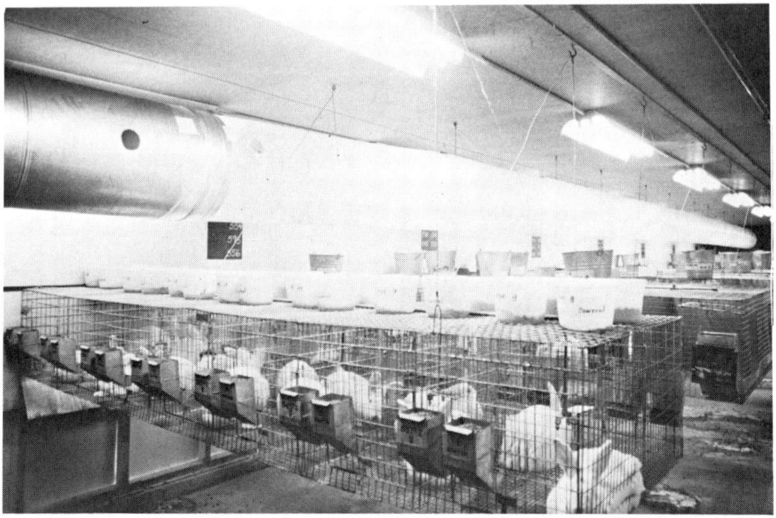

Hutches are arranged in rows for easy feeding and other care.

Detail of hutches

The floor area needed to house a mature rabbit in an individual compartment varies with the breed and size. Hutches generally are built 30 inches wide and should never be less than 24 inches wide. This enables the handler to easily reach to the rear of the hutch. The length of the hutch should vary with the breed:

Small breeds	2 ft. long
Medium breeds	3 ft. long
Giant breeds	4 ft. long

This amount of space accommodates both the doe and litter until the young rabbits are removed at about 8 weeks. An 8-week-old litter may be placed in a hutch and fattened with pellets and water until 10 weeks of age to get maximum weight and a better pelt.

A free-standing outdoor hutch should be covered with a tight sloping roof 2 feet above the floor in the front and 18 inches above the floor in the back. A 6 to 8-inch over-shot roof in front is satisfactory for shade.

The floor is the most important part in the construction of the hutch. It should be self-cleaning to safeguard the health of the rabbits. Floors can be made of 16-gage wire or perforated galvanized metal.

Whether you arrange the hutches in single, double, or triple tiers depends upon how much room is available. If you have enough room, waist-high, single-tier hutches are preferable as they are most convenient for observing the rabbits and will also save time and labor in feeding and management.

The two- or three-tier hutches, necessary when space is limited, are not entirely satisfactory for caring for and observing the animals in the bottom and top tiers. The inconvenience of squatting or stooping to feed and care for rabbits in the bottom tier and of having to use a stool or ladder for the top row of a three-tier arrangement results in additional labor and time as compared to a single-tier arrangement. Rabbits in top and bottom hutches often get less attention than those at eye-level.

Rabbits are more easily cared for in well-built hutches than in poorly constructed temporary ones. Self-cleaning all-wire hutches need no bedding and you can easily keep them in good condition.

Keep clean-up in mind

When we speak of "self-cleaning" cages, we are talking about urine and feces dropping to the floor or ground. If your hutch is located outside, droppings can fall to the soil. Urine will soak into the ground. In a garage, or some other building with a concrete or some other solid floor, you will need to put in a metal pan to catch the droppings. Or you can use a sheet of plastic inside a cardboard box. A shed or barn with a dirt floor offers another possibility.

Wire hutches for doe and litter suspended from the ceiling

How to build a hutch

A state university-developed plan suitable for four rabbits is shown on the following pages. It has four hutches with a V-shaped self-feeder serving all four in the middle. You may want to stretch the dimensions a bit to make each hutch 30 inches wide and 36 inches long.

Required materials

4 pieces 2" x 4" x 36"
3 pieces 2" x 2" x 87½"
2 pieces 2" x 2" x 46½"
2 pieces 2" x 2" x 20⅜"
8 pieces 2" x 2" x 32"
8 pieces 2" x 2" x 20"
4 pieces 2" x 2" x 24"
(for diagonal braces)
8 pieces 1" x 2" x 9½"
12 pieces 1" x 2" x 23"
8 pieces 1" x 2" x 20¼"
1 piece 1" x 2" x 23½"
8 pieces 1" x 2" x 22"
4 L brackets
12 hinges
6 safety latches
28 feet of 24" wide mesh ½" x 1"
12 feet of 24" wide mesh 2" x 1"
2 pieces 4' x 8' plywood ½" thick
Nails 1" long, 2" long, and 3" long
Shingles or other roofing material if hutch is to be used outdoors.
2 boxes of fencing staples

The upper assembly is built separately and then placed onto the frame. This hutch has legs 3 feet high, and you may want to place them into the ground a few inches for added stability. The braces must be miter-cut to either 45°, 60°, or 30° angles, depending on their use.

This is a hutch built to stand on its own in a sheltered back-yard location. It will need shade for summer, particularly in hot climates. And you should include roll-up canvas siding to protect the rabbits from cold winds and rain.

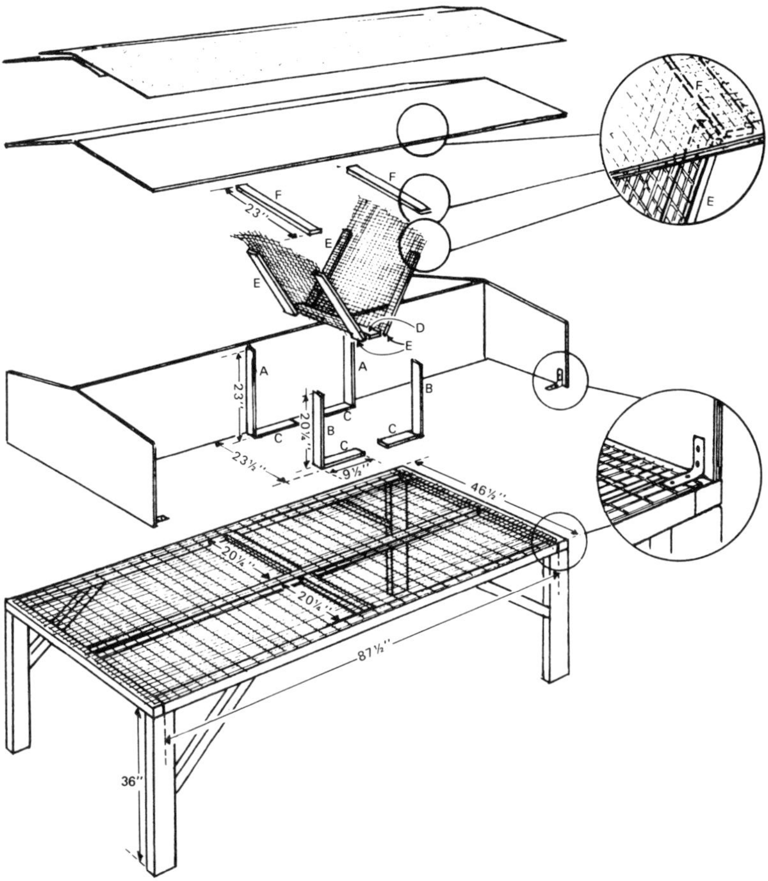
F
F
23''
E
E
D
E
A
A
B
23''
C
B
C
C
C
20¼''
23½''
9½''
46½''
20¼''
20¼''
87½''
36''

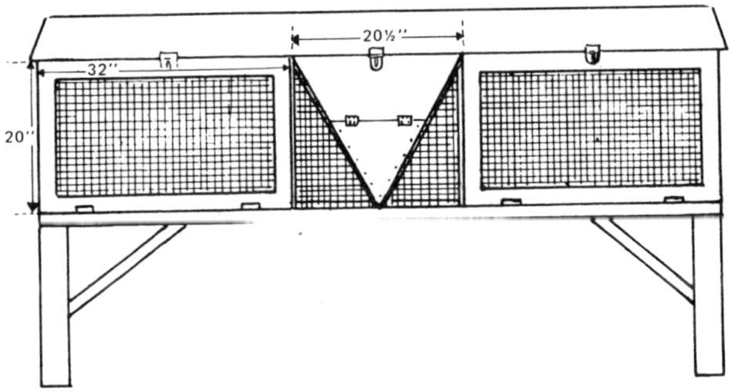
20½''
32''
20''

Build your own cages

Another option for the handyman is construction of an all-wire hutch. Buy 1 x 2 x 18 inch rolls of wire for the job. A 50-ft. section should be enough to make four hutches 30 x 36 x 18 inches. Select ½ x 1 x 30 inch wire for the floors.

Some does have a distressing habit of kindling on the wire instead of in the nest box. Some of the very small newly born, crawl through the larger side wire and fall to the floor. Or they may get their heads caught. A strip of the smaller mesh, about 4 inches high around the sides at the bottom of the cage can help solve this problem.

Some cages have top openings; others have side openings. Most breeders now prefer the side opening. It sometimes is difficult to lift rabbits out of the top. Feeding and watering are awkward.

It may be better to buy

Before you spend any money on materials, consider buying a hutch. Price them in a Sears catalog or through other rabbit supply channels. You may find a professionally designed and built unit may be better in the long run.

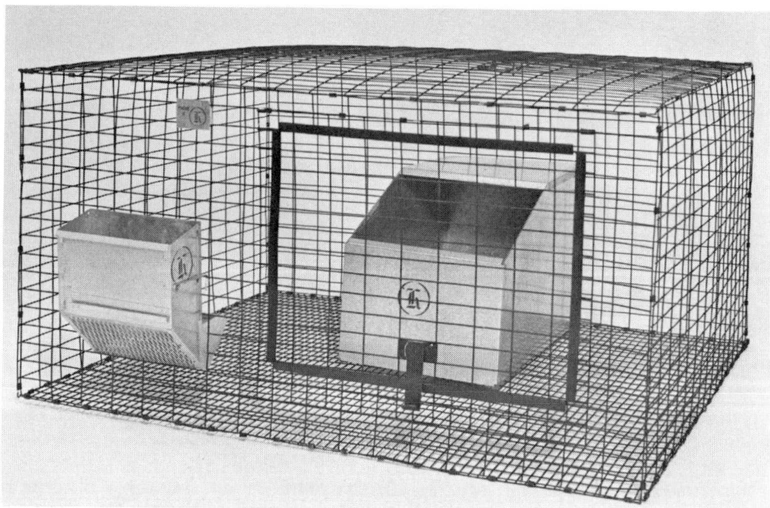

A manufactured rabbit hutch

The unit shown here comes in a variety of sizes. You can order a 24 x 36 inch unit for the smaller breed doe and litter, or you can go to 30 x 36 inches for the larger breeds of rabbits.

The floor is heavy 16-gauge ½ x 1 inch mesh; top and sides have 2 x 1 inch mesh. Sharp wire ends are protected by plastic.

The side mounted door opens inward and can be latched to the top to make it easy to work with your hands inside the hutch. Self feeder attaches to the sides. The next box fits inside.

This hutch, and those made by other companies, usually are shipped knocked down. Hardware and instructions for construction are included.

Typically, wire hutches are suspended from the ceilings or rafters of a building. They are mounted end-to-end to form an aisle through the building. This simplifies feeding, watering and care of the rabbits. You just walk down the aisle, servicing each cage in turn. Most breeders use a cart to carry supplies as they move along the row. A smooth hard surface makes the work go faster. Pave the ground in front of the cage with concrete or blacktop. Leave the soil beneath the cages to help soak up urine.

Pellets are fed by most rabbit producers

Feeders

Your feeding system should be geared to efficient use of commercial pelleted feeds. While some breeders still feed clover or alfalfa hay and an occasional garden green, the great majority now depend upon pellets for the complete ration. Self-feeding as much of the ration as the rabbits want is the accepted way to make sure they get what is necessary for fast, profitable growth.

When you are starting out, and have a small number of rabbits, a simple crock or can may do the job. But as numbers grow, select a commercially made metal self-feeder. These units save you time and work, protect the feed and provide easy access for the rabbits. Less waste of feed will pay for the self-feeder in a short time.

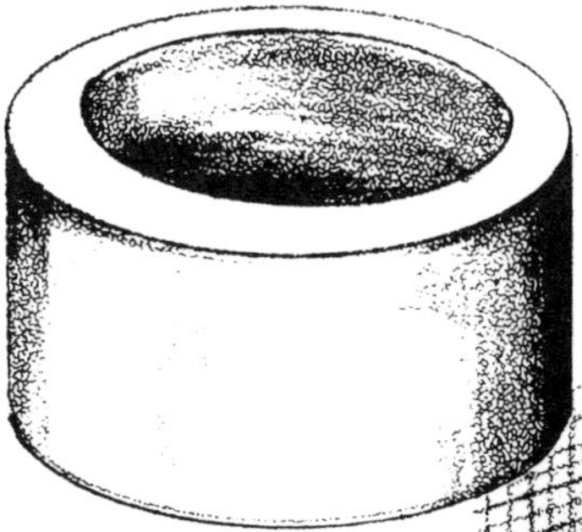

A heavy porcelain crock.

A metal feeder.

Crocks and Cans

Crocks make a good container for feed. These ceramic units with a wide lip are made for rabbits and are available in pet or rabbit supply stores. You may find similar units in a hardware store. These are not easily tipped and the lipped edge prevents feed waste.

Some breeders use a cut-down coffee can for a feeder. Cut the can down to about two inches, two-thirds of the way around the can. The remaining, full-length part of the can is used to anchor the unit to the cage. Fold over the sharp edge of the can on the short side to keep it from cutting the rabbits as they feed. Wire or clip the high side to the cage.

Young rabbits are not very tidy. They have a bad habit of climbing into an open feed container where they get feed dirty with manure or urine. Once this happens, rabbits won't eat the feed. And if they do, it can lead to serious health problems. This is why most producers prefer the metal self-feeder.

Metal self-feeder for rabbits

This self-feeder is typical of those available through rabbit supply houses. The unit clips to the side of the cage. The trough is inside where rabbits can reach the feed easily and is wide enough so that several rabbits can eat at one time. It provides a continual supply of feed while eliminating feed scratching and waste. The rabbits can't foul the feed with urine or droppings.

The feeder is filled from the top, outside the cage. You just walk down the cage row putting a supply of feed into one cage after the other. The units hold enough so that you don't have to feed every day.

Sifter-type self-feeder

Another version of the self-feeder has a screened bottom. As feed is put in, unappetizing feed fines and dust drop out. The screen is small enough to keep pellets inside but lets the less desirable portion fall out. Feeding arrangement inside the cage is the same as the solid bottom self-feeding unit.

Sifter-type self-feeder lets fines fall though.

Another way to feed rabbits is through a cage cup which is clipped to the side of the cage. It's an all-metal unit that is easily cleaned. It's size and shape helps solve the problem of contamination by rabbits.

Hanging cage cup

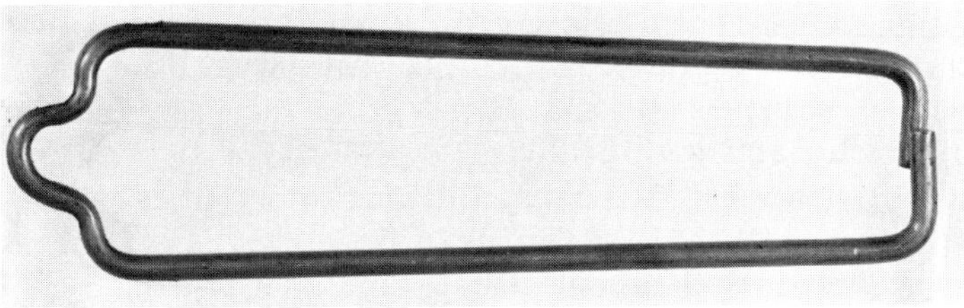

Salt spool Hanger

Rabbits need extra salt. One good way to supply it is with a salt spool hanger. This aluminum unit hangs from the top of the cage and keeps the salt clean and dry within reach of the rabbits. Spooled salt, available in feed supply or pet stores, has a hole in the center. It is fitted into the bottom of the hanger.

Creed Feeder for Young

Young rabbits need plenty of feed to reach 4 pounds in just 8 weeks. Many producers use a self-feeder to make sure the bunnies get a head start. This unit has a porch arrangement that lets small rabbits in and keeps the doe out. It fastens to the floor of the cage with clips. Cover is removed to fill with feed. Use special creep feed pellets for young bunnies.

Feeder for Hay

If you decide to feed hay, build a wire rack that attaches to the outside of the cage. Wire hay racks with sloping sides and narrowing at the bottom, seem to be preferred by leading rabbit growers for feeding the hay portion of the diet.

One-inch 14-gage wire serves this purpose. Odd pieces of wire may be used even if the hutch holes are slightly larger or smaller. However, if the hutch holes are rather large, the hay will be pulled out too rapidly and will become soiled and wasted, at the same time interferring with the self-cleaning of the floor.

Keep in mind that rabbits have nothing to do but eat and grow, hence rather small nest wire permits sufficient hay stems to be pulled through. The rabbit chews only one stem at a time, beginning with one end instead of several stems crosswise, unlike larger farm animals.

A narrow trough under this V-shaped hay rack will catch and save the valuable fine leaves and blossoms of the hay. This trough should be narrow and have numerous small partitions to discourage the rabbits from getting into the trough lengthwise and soiling the contents. These partitions may be made of wood or of wire. An added convenience is to construct this small trough so that it may be pulled out like a drawer for cleaning and disinfecting. The bottom of the trough may rest on the floor of the hutch. The sides of the trough may be two to three inches high, or whatever height seems necessary to prevent manure and urine from getting into the trough. If edges are bent inward the rabbits will be less able to scratch out food materials.

A two-compartment hutch on the same floor level, such as the one shown on page 49 can have the hay rack in the center to separate the compartments and permit the occupants of both to eat from the sides of the same rack, thus saving time and labor for the attendant. The rack may be filled from the end by simply pushing the hay into the end of the rack without having to open any hutch doors. The rack should extend to the top of the hutch or be protected so that no rabbit can enter it and it should be large enough for a 24-hour supply of hay.

Rabbits need lots of water

Rabbits need clean, fresh water 24 hours a day, every day. Water may be even more important than feed for a fast-growing litter. A doe and her litter will drink about one gallon of water each day in the summer. It's a real challenge to supply this amount of water on a hand-chore basis.

When you start out, you probably will want to be as economical as possible in supplying water. The same type of crock that is used for feed also can be used for water. In hot weather, it will have to be filled twice a day to supply that gallon to the litter.

Cans also can be used. They tend to rust and are easily tipped over. Coffee cans are especially useful during cold weather. Ice can be broken and removed from the cans.

Crocks and cans offer the same sanitation problems we mentioned for feed, when they are used for water.

Young rabbits eat, drink and fiddle around. They climb into the water, and use it as a means to keep cool. Water becomes contaminated with urine and feces. Cleaning is a never ending job.

Sipper tube in a soda bottle.

One way to solve the watering problem is with a sipper tube. Rabbits quickly learn to sip water from the tube. Buy a kit that includes a stopper that fits a large sode bottle and the necessary tubing. It hangs on the outside of the cage with the tube pointed inward.

While this unit solves the contamination problem, filling it is a chore, particularly when you have a large number of rabbits. The bottle will break if left filled with water in below-freezing weather.

Go to automatic watering

The first mechanization you install in the rabbitry should be an automatic watering system. It is far superior to water crocks or coffee cans. An automatic system eliminates the tedious and time-consuming chores of washing, disinfecting, rinsing, and filling. It will supply fresh, clean water for the rabbits at all times. When an automatic watering system is properly installed, dirt and fur will not collect in it and plug it up.

Install electric heating cable inside the pipe for use in cold weather. It will prevent freezing so that the rabbits can sip water at anytime, day or night.

Labor saving dew drop valve

The key to success of the automatic watering system is the ability of a rabbit to open the valve and take a drink. The rabbit gets water by licking or biting the plunger end. Rabbits learn to use the system easily, even young just out of the nest box. The unit operates on very low pressure of 3 to 5 pounds per square inch.

If you can cut and thread pipe, you can install an automatic watering system. Conventional systems sold by rabbit and poultry supply houses consist of a pressure-reducing tank equipped with a float valve, a ½-inch supply pipe, a watering unit for each hutch, and valves. The valves are used to bleed out air bubbles, to drain the system as needed, or to shut off the water.

One-gallon tanks or smaller tanks sometimes are used where the weather is warm. Such tanks are emptied more often. The constant flow of water in and out of the tank keeps fresh cool water before the rabbits at all times.

Place supply tank high

Install the pressure tank 1 foot or more above the highest hutch. If the supply pipe is raised to clear the feeding alleys, then install the tank about 1 foot above this highest point.

Raised supply pipes may require vent pipes to keep air bubbles out of the system. Install the vent pipe at the highest point in the supply line. See that the open end is at least 1 foot above the water level in the tank. If it is necessary to change the level of the supply line from one row of hutches to another, use a piece of rubber hose to make the connection.

Determine the correct height for the tank by fastening a rubber hose to the tank outlet and then to the supply pipe. Raise or lower the tank until the valves, or dewdrops, from which the rabbits drink, have the proper tension. If there is too much tension or pressure on the valves, the rabbits will not be able to trip them. Under too little tension, the valves will drip.

Where to locate valves

The proper height for the water valve is 9 inches from the hutch floor for medium and heavy breeds and 7 inches for the smaller breeds. The pipe may be hung on the outside and at the back of the hutch so no water drips on the rabbits and the hutch floor. An opening in the back of the hutch will permit the rabbit to use the valve.

When hutches are back-to-back, use one pipe for supplying water to both hutches. Use a four-way outlet and short nipples for installing the valves.

You can install one drinking valve for each hutch by drilling and tapping the supply pipe and screwing the valve into it.

If you are not equipped to make the plumbing installation, substitute a ¾-inch rubber hose for the ½-inch supply pipe. Cut a hole in the hose and screw in the valve. Plastic pipe may be used in a similar manner.

Check the automatic watering system periodically, especially when you put a rabbit in a hutch that has been unoccupied for several days. When valves are not used—even for a few days—minerals in the water may cause them to stick.

Rabbits need a warm, snug nest

Does like a quiet, secluded place to give birth. Rabbits are born hairless, blind and unable to do much more than wiggle to a teat to nurse. They open their eyes at 10 or 11 days. The litter needs the protection of a good nest box in the first critical weeks.

Other things needed in a nest box are good ventilation and easy cleaning. It needs to be designed to keep the rabbits in when they are very young and still let them out easily when they are old enough to go to the feeders and waterers for themselves.

The nest may have a removable partial cover. This makes it easy to check on the litter and to change the bedding or remove a liner when necessary. Covers are seldom used in hot climates.

Wood has excellent insulating qualities and makes a good nest. Most rabbit producers build them of pine since it is light, low cost and easy to work with. However, rabbits can gnaw on soft pine and exposed edges should be covered with strips of metal.

The box should be 12, 16, or up to 20 inches, depending upon the size of the rabbit breed. Plans for wooden nest boxes of three different shapes are shown here. Select one that will go through the door of the hutch. Remove them for thorough cleaning between litters.

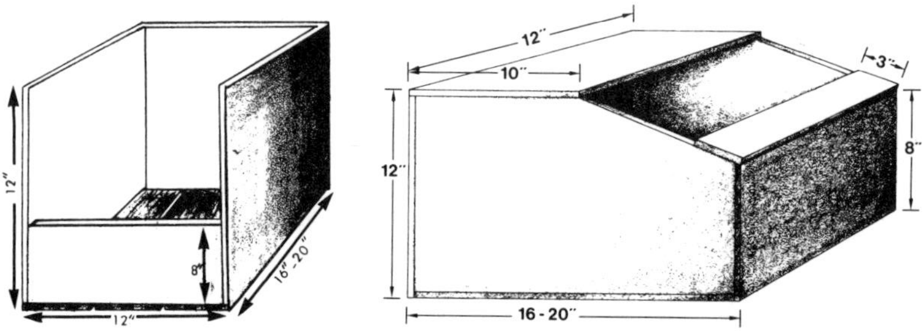

Build a wooden nest box.

A commercial metal nest. A warm floor can be added

Choose a metal nest box

This commercial nest box is designed for comfort and easy sanitation. Top and sides are galvanized metal. The floor is wood which provides a warm base for the nest. It is removable for cleaning. This nest is built in 10 x 18 inch sizes for small breeds and 11 x 22 for the larger breeds.

Many growers in hot climates are using wire nests built of the same materials as hutches. These permit cooling during the hottest time of the year.

Some feature a hardboard bottom to serve as the base of the nest. Others use a cardboard liner. The cardboard is cut down in summer. One model has a waxed cardboard liner that is thrown away after use by one litter. Never use cardboard alone...it just won't stand up to the wear and tear given by a litter of rabbits.

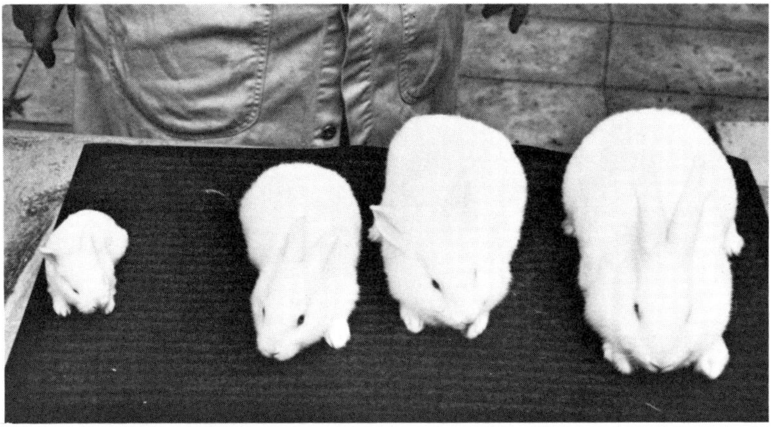

These rabbits are two, four, six and eight weeks old.

Chapter 5

Managing breeding for high production

Your role as a rabbitry manager begins the day you bring your first rabbits home. Getting them started correctly will help you get off on the right foot in your new enterprise. Make sure things go smoothly for you and for your rabbits.

Remember that the placid domestic rabbit is not too far removed from the wild. Handle them gently in transit. Avoid extremes in temperature, particularly too much heat.

Be ready for them when you get them to the rabbitry. That means hutches prepared and a good supply of food and water. Ask the former owner what he had been feeding the rabbits. Try to keep them on the same brand of feed so that the transition will be as easy as possible.

Rabbits are easily excited. Always talk, sing, or make some noise as you enter your rabbitry to let them know you are coming. Do not tiptoe in your rabbitry or protect the rabbits from natural noises. Otherwise, every strange noise will cause them to stampede. The use of a radio playing music sometimes can help keep your rabbits calm.

Handle your rabbits regularly to prepare them for movement in and out of hutches. It is important that they do not struggle when you weigh them, breed them or check for pregnancy. Rabbits can scratch. If one does, it is usually because you have picked it up or tried to handle it in such a way that the animal feels insecure and tries to free itself.

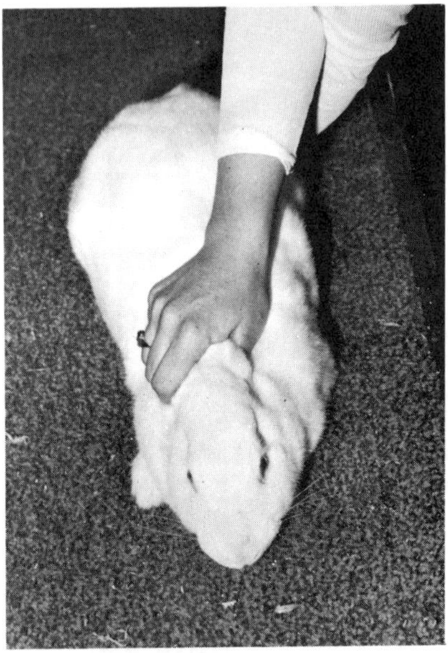

Using the right hand, grasp the fold of skin over the rabbits shoulder.

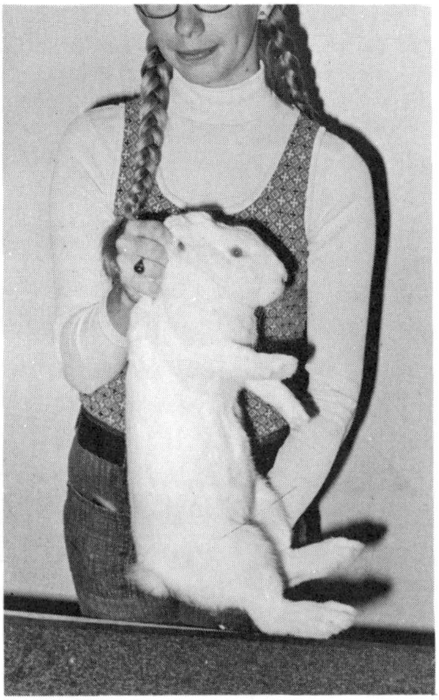

Support the rabbit by putting the left hand under his rump.

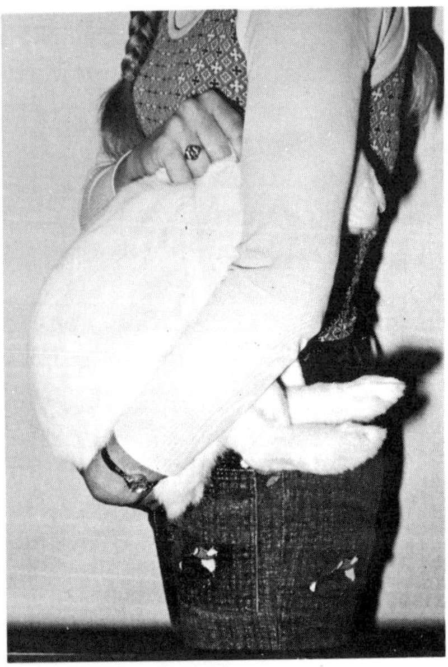

To carry the rabbit a distance, tuck it under your arm near your body. Retain neck and rump hold.

How to pick up a rabbit

A 10-pound rabbit can be hard to handle. Develop sound techniques in lifting and carrying rabbits. The ears aren't handles...never use them in lifting.

You can lift and comfortably carry small rabbits by grasping the loin region gently and firmly. Put the heel of the hand toward the tail of the animal. This method prevents bruising the carcass or damaging the pelt.

To lift and carry a medium-weight rabbit, let the right hand grasp the fold of skin over the rabbit's shoulder. Support the rabbit by placing the left hand under his rump. As you pick up the head, reach over with the other hand and grasp the rump. Lift the entire body of the rabbit at one time so the rabbit feels secure. Use the same technique for turning a rabbit over. To carry the rabbit for a distance, pick it up and tuck it under your arm near your body.

Never handle some breeds, such as the Rex and the Tan, over the top of the shoulder because of possible damage to the fur. Grasp them gently but firmly by the flesh on the side and support with the other hand under the belly.

Housing arrangements

The buck should have a hutch of his own, away from does. Never mix does and bucks except for breeding. Each doe also must have her own hutch with room to raise her litter. One buck per 10 does is considered a good ratio for small operations. Large commercial rabbitries average one buck for 20 does.

The age for first mating

The age for first mating depends upon the breed of rabbit and its individual development. Some rabbits of the same breed take longer to mature. Others are ready so soon you have "teen-age" pregnancies.

Smaller breeds develop more rapidly and are sexually mature at a much younger age than medium-weight or giant breeds. Does should be mated when they reach maturity; some difficulty may be experienced if mating is too long delayed.

On the average, the smaller breeds may be bred when the bucks and does are 4 to 5 months old, the medium-weight breeds at 5 to 6 months, and the giant breeds at 8 to 10 months.

Some individual rabbits within a breed develop more rapidly than others, and does usually mature earlier than bucks. In commercial production, it is the general practice to hold bucks a month longer than does before breeding for the first time.

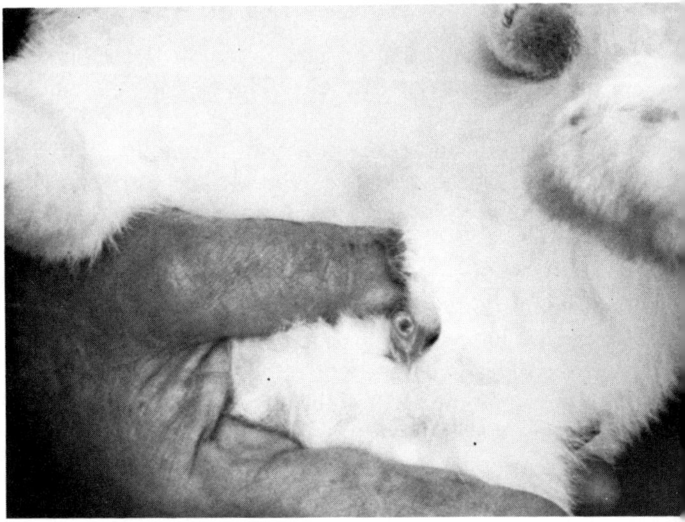

Buck—the sex organ appears as a rounded tip.

Which is which?

It's easy to tell male from female in most species of the animal world, but not in rabbits. It takes an expert to look at two rabbits in a pen and determine which is the buck and which is the doe without an examination of the reproductive organs.

There's an established technique for this:

To keep the rabbit from struggling, restrain it firmly, yet gently. A commonly used method is to hold the rabbit on its back between your legs with the head up. With your left hand restrain the rabbit around the chest holding the front legs forward alongside the head.

Using the right hand, place the thumb behind the right hind leg and use the index and forefinger to depress the tail backward and downward. The thumb is then used to gently depress the area in front of the sex organs to expose the reddish mucous membrane. In the buck, the organ will protrude as a rounded tip, while in the doe the membrane will protrude to form a slit with a depression at the end next to the anus.

When does are ready

Rabbits do not show regular estrous cycles, that is, recurrent periods of sexual desire. During the breeding season the doe remains in heat

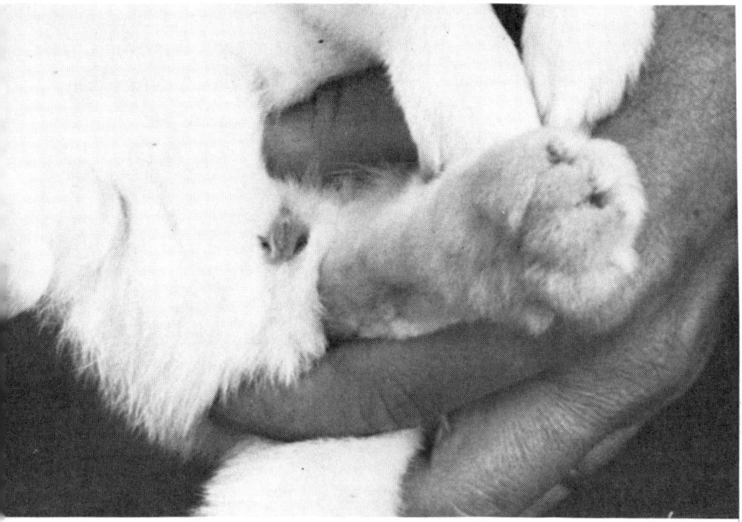

Doe—the sex organ appears as a slit.

for long periods of time. If she is not bred, the follicles in the ovary remain large and active for a period of 12 to 16 days. After this time they begin to regress. Meanwhile, new follicles grow to replace them. As a result, active follicles are present at all times during the breeding season. There may be a transitional period, while the new set of follicles is growing and the old set is retrogressing, when the doe lacks interest in the male and is temporarily sterile.

Breeding technique

There are many breeding techniques used, but the natural mating method is recommended. Check does regularly. A doe is at the peak for the best conception when the vulva is moist and a bright pink to a reddish color all the way to the tip. As the cycle is waning, the vaginal opening becomes bright purple.

Always take the doe to the buck's cage. If you take the buck to the doe's cage, she may fight or resist him. The female usually accepts the buck within a few minutes. After the doe accepts the buck, he falls over to the side or even backwards. Within a minute or so he'll be back up to repeat the mating. Watch to see that the buck services the doe. If the doe urinates immediately, she may lose the sperm. If that happens, take her back right away to be re-bred. Also be sure that the buck does not lose his sperm before completing the act of breeding. You can lose a month's production if you do not watch the breeding. Return the doe to the buck's cage for re-breeding 6 to 12 hours after the first mating.

Bucks should not be used more often than once every 3 days. However, daily breedings for short periods are satisfactory. Keep bucks where they cannot see the doe.

Sometimes, the doe will not allow the buck near her. When this happens, try again in one or two days. When mating occurs, return the doe to her hutch and record the date and name of both doe and buck. This is important. Many litters have been lost because the date of the breeding was not recorded and the nest box was not placed in the pen by the day of kindling (birth).

If you house a shy or hard-to-breed doe in a cage next to the buck for 12 to 24 hours before breeding, she may prove easier to breed. If a doe continues to be difficult to mate, consider culling her.

A first-time buck may be slow about breeding. Be sure the doe does not scare him or harm him. Do not breed your bucks until they are near maturity—breed lightweights at 4 to 6 months; heavyweights at 5 to 7 months.

Force matings are sometimes needed

It is difficult to get some does to accept service. Such does may be restrained for mating. Use your right hand to hold the ears and a fold of the skin over the shoulders, and place the left hand under the body and between the hind legs. Place the thumb on the right side of the vulva, the index finger on the left side (you may prefer to use the index and second finger), and push the skin gently backward. This procedure throws the tail up over the back. Support the weight of the body by the left hand, and elevate the hindquarters to the normal height for service.

Bucks and does accustomed to being handled will not object to such assistance. It is well to hold the doe in this way the first few times a young buck is used. This will expedite matings and insure service in difficult cases.

With a little patience and practice you can develop this technique to procure nearly 100-percent matings. This does not necessarily mean that all of the restrained will kindle, but the technique will help in increasing the number of kindlings.

Whether it is worthwhile to force-breed for increased pregnancies depends largely upon the number of does in production. Owners of large commercial rabbitries generally do not force-mate their does due to the increased labor and time involved. For small rabbitries, a few extra litters could be worth the effort, and for breeders of pedigreed show stock, where animals are removed from production part of the year, forced breeding will help maintain a supply of replacements and stock for sale.

Colony breeding plan

Colony breeding is sometimes used with hard-to-breed does. Leave the doe with the buck in his cage for 28 days and then return her to her cage. Immediately put a nest box in the cage. This method ties up a buck for 28 days. Some breeders put 8 to 10 does in a ground-level pen with a buck until they are bred. Problem: you never know when a doe is due to kindle.

Summer can be difficult

There are seasons of the year when breeding is very difficult. During summer heat, most older bucks, especially the heavy-weights, become temporarily sterile and you need the services of young bucks about 6 to 8 months of age. In the fall and early winter, the does may not reproduce. You need patience and persistence to get your does into production.

During the summer and early fall months, do not over-feed rabbits. Overly fat animals are hard to breed and may have difficulty, both during gestation period and at kindling time. Females sometimes get so fat the birth canal is restricted or even closed. Limited feeding during pregnancy is a wise move. Bucks also should be kept trim through restrictive feeding.

When to rebreed a doe

Does customarily are rebred before they wean their litter. Prompt rebreeding makes it possible to raise five or more litters per year.

Commercial breeders may rebreed does 14 days after kindling. They also use multiples of seven days, such as 28, 35, 42 and 56 days.

One of the most popular systems is the five week schedule, or 35 days. In this plan, the litter is weaned at 56 days, or 21 days after the doe has been rebred. This allows the doe nine days of rest before she kindles again. This helps keep her in good physical condition.

Eight weeks (56 days) is the normal weaning time. If a doe is to be bred on an earlier schedule, the litter must be weaned at a younger age. For example, rebreeding at 14 days would require weaning at four weeks of age. Separate quarters for the young rabbits will be needed when this plan is used.

Breeding at 42 days makes it possible to raise five litters per year. That includes five gestations of roughly 150 days total, and 175 days of interval between kindling and breeding, for a total of 325 days out of the year. There are a few extra days here that could make your production around 5½ litters annually. However, enough breeding periods are likely to be missed, particularly in hot weather, to hold the average to around five litters.

How to time breeding and weaning

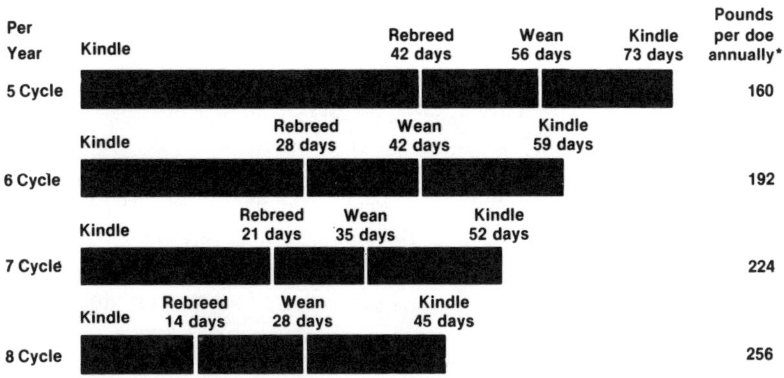

All bunnies marketed at 56 days.
* Average 8 per litter, sold at 4 lbs.

The Purina Research Farm has been a leader in stepping up the production cycle for rabbits. Over the years, they have selected does that can produce eight litters per year. The chart above shows the timing of 5, 6, 7 and 8 cycles per year. The pounds of rabbit produced per doe with an average of eight per litter climbs from 160 to 256 per year. This requires a higher-priced, top quality ration and more labor, but the investment in equipment and breeding stock remains the same.

The Purina Research Farm also has kept a close eye on feed efficiency. Many litters have been raised on 2.6 to 2.8 pounds of feed per pound of gain. In a few tests, the goal of two pounds of feed per pound of gain has been obtained. This is a giant step forward and shows what may be accomplished in the future.

Which breeding system?

Select a breeding system and carefully choose your matings to make it work. There are three general programs that are sometimes used.

Outbreeding: This is selection of stock with different parentage. Closely related relatives never are mated. This requires bringing in new bucks to keep fresh genetics flowing into the herd. This is the system, when combined with selection on the basis of performance, that will bring fastest progress for you.

Inbreeding: This is the mating of animals which are closely related. In its strict interpretation, it is the mating of sire to daughter, son to dam, or brother to sister. Such mating is not harmful in itself but it is a sure, rapid, and effective means of bringing out certain good and bad characteristics. It will intensify poor qualities as well as good qualities. Many rabbit breeders feel that after inbreeding once, it is best to then introduce unrelated stock of the same breed. This system is not recommended for beginners. Breeding unrelated males and females gives better over-all results. Inbreeding should be done at the risk of mass failure and under the supervision of an experienced geneticist. If you do practice inbreeding, observe all young rabbits closely all their productive life and get rid of all rabbits with poor characteristics.

Cross-Breeding: This is the practice of mating a purebred rabbit of one breed with a purebred rabbit of another breed. This plan of breeding is valuable in the production of new varieties and should be attempted only by breeders of considerable experience. A breeder walks a fine line between heterosis and prepotency. It is difficult to emphasize good traits unless you keep records and breed animals that produce superior offspring, or you can inbreed and kill off the ones with undesirable traits. Remember that animals that are alike do not necessarily begat like offspring. So keep performance records on young from every mating in which you seek improvement in genetics.

Each breed has characteristics that have been acquired through breeding and selection, which distinguishes it from all other breeds. When two animals are bred, the characteristics most likely to appear in the offspring are those common to both parents. The more parents differ, the more prominent will be the unimproved characteristics in the offspring. It is common to find some older commercial breeders of rabbits with a history on all rabbits mated to a dozen or more generations.

Line breeding: This is the same in principle as inbreeding, except matings are made with animals of the same families a few generations back. As a result, characteristics of mated individuals, whether desirable or undesirable, are not found in the offspring as rapidly as when inbreeding is practiced. Probably most attempts by the beginner at inbreeding or line breeding are made to avoid buying a new buck. Rather than take chances of obtaining inferior offspring, by making close matings, it would be better for the beginner to buy a new buck of the desired type, when it is necessary to breed does that are related to the herd buck. The type and condition of the individual are just as important as the pedigree.

Palpating shows if she's pregnant

An accurate method of determining pregnancy is important in managing the breeding herd. Test-mating (placing the doe in the buck's hutch periodically) is not accurate because some does will accept service when pregnant and others will refuse service when not pregnant. Diagnosing pregnancy by the development of the abdominal region and the gain in flesh is not dependable until late in the pregnancy period.

Pregnancy can be quickly and accurately determined by palpating the doe 12 to 14 days after mating. Restrain the doe for palpating. Hold the ears and the folds of skin over her shoulders in the right hand; place the left hand between the hind legs, slightly in front of the pelvis; place the thumb on the right side and the fingers on the left side of the abdomen; exert light pressure, and move the fingers and thumb gently backward and forward.

If the doe is pregnant, you should be able to distinguish the embryos as marble-shaped forms that slip between the thumb and fingers. Handle the doe gently, and use only light pressure on the abdominal cavity.

Give does diagnosed as pregnant, all the feed they will eat. Non-pregnant does should be rebred and kept on a maintenance ration until they are diagnosed pregnant.

Accurate determination of pregnancy by palpation takes practice. If you are inexperienced, repalpate the does diagnosed as non-pregnant, one week later. If a mistake has been made in the first palpation, the doe can be put on a full ration and provided with a nest box at the proper kindling time. With practice, you may develop enough skill to diagnose pregnancy as early as the 10th or 11th days. The photos at right show how to palpate.

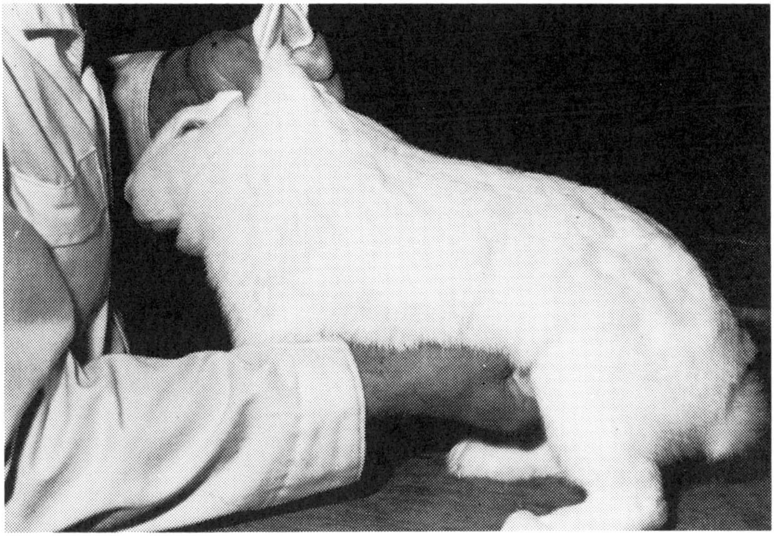

To determine pregnancy, palpate the doe 12 to 14 days after breeding. Restrain the doe with the left hand holding next fold and ears.

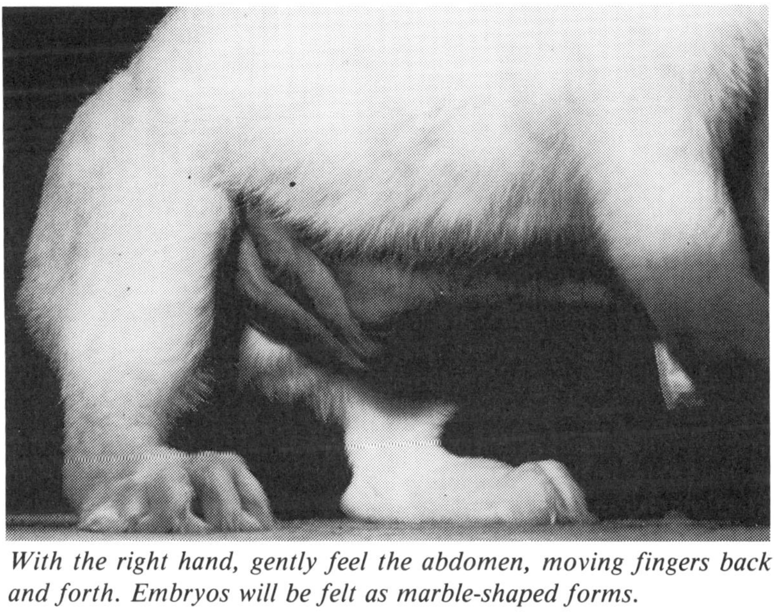

With the right hand, gently feel the abdomen, moving fingers back and forth. Embryos will be felt as marble-shaped forms.

Be ready for kindling

"Having babies" in the rabbit world is called kindling. The normal gestation period is 28 to 32 days. This date comes around in a hurry and you must keep records of the breeding date so you will have a nest prepared for the doe.

Start the litter off right, in a clean, sanitized nest box. Don't expose them to the problems of the past.

Don't under-estimate the importance of disinfecting the nest box. To completely ready the nest box, you'll need to disinfect and clean the entire box thoroughly. Don't forget to scrub down boxes between litters. You'll need a 2-pound can full of wood shavings to place in the bottom of the nest. Next, you'll need plenty of good, clean straw. The doe will make her own nest. If the doe eats a lot of the straw, don't be alarmed, just add more straw. If all of your breeding herd does are eating a lot of straw, re-check your feeding program to see if something is missing.

Other materials that make a warm, absorbent base, can be used. Servell, a commercial product made from shredded sugar cane, is preferred by commercial producers. It is sold in 100 pound bales. Many breeders put in a disposable cardboard liner, then add bedding. The doe will top off the material you provide with fur pulled from her body.

Sometimes does fail to pull fur to cover their litter, or they kindle the litter on the hutch floor and let them become chilled. If you discover the young in time, you may be able to save them by gradual warming, even if they appear lifeless. Arrange the bedding material to make a comfortable nest and place the warmed young into it. The doe usually will take over from there.

The doe's fur is easily removed at kindling time, and you can pull enough from the doe's body to cover the litter in the nest. Keep extra fur on hand for such situations. Remove some fur from nests where does have pulled out an excessive amount and keep it handy in a bag or box so it will remain clean. You don't have to sterilize or deodorize the fur or take any special measures to prevent the doe from smelling the strange fur.

A day or two before kindling, the doe usually consumes less food than normal. Do not disturb her, but make her as comfortable as possible. You may tempt her at that time with small quantities of green feed. This will have a beneficial effect on her digestive system.

Most litters are kindled at night. After kindling, the doe may be restless. Do not disturb her until she has quieted down.

Kindling-time complications

Anterior or breech presentation of young at birth is normal. If the doe is in proper condition for kindling, complications are rare. Pregnancy, however, makes a heavy demand on the doe and lowers her vitality, making her more susceptible to disease. A few days before or several days following kindling, pneumonia may develop. To treat pneumonia, you must detect it early. The doe's head is held high and tilted backwards. Breathing is difficult. Make the doe comfortable, and add a little green feed to the ration. Antibiotic injections can treat colds and pneumonia, reducing mortality.

Caked breast may be caused by milk not being removed from the breast or by injuries. Early symptoms of caked breast are firm, pink breasts that feel feverish to the touch. As caked breast develops, the tissues around the involved teats become enlarged and hard. The skin turns dark, the ends of the teats become discolored and tender, and the doe refuses to allow the young to nurse. Rub lanolin on the teats, and massage the involved portion of the breast. You may restrain the doe to allow her own young or those from other litters to remove the milk. You may also strip the milk from the teats, taking care not to use too much pressure. Do NOT lance the tissues.

Mastitis or "blue breast" is caused by bacterial infection and may be very cantagious. The doe fails to consume her feed and is inactive. The breast is congested and feverish, turns red or purple, and the teats are discolored. Reduced feed, some green feed, and antibiotic injections are the best known treatment.

Care of the young litter

On the day of kindling, or soon after, inspect the litter and remove any deformed, under-sized, or dead young. If you are careful and quiet during the inspection, the doe usually will not object. There is no danger of causing her to disown the young. If she is nervous and irritable, place some tempting green feed in the hutch immediately after inspection to distract and quiet her.

Litters vary in size. The utility breeds usually average eight young. Some may number 12 to 18. For commercial purposes, 7, 8, or 9 may be left with the doe. Does from strains that have been developed for heavy production, may care for 9 or 10.

You can transfer some of the baby rabbits from a large litter to a foster mother who has a small litter. Adjusting the number of young to the capacity of the doe insures more uniform development and finish at weaning time. Mate several does at the same time so that they will kindle together. For best results, the young that are transferred should be within 3 or 4 days of the age of the foster mother's young.

Causes of losses in newborn litters

If the doe is disturbed, she may kindle on the hutch floor, and the young may die from exposure. Even if predators—cats, snakes, rats, weasels, minks, dogs—cannot gain access to the rabbitry, they may be close enough for the doe to detect their presence. She may become frightened and kindle prematurely. If she is disturbed after the litter is born and jumps into the nest box, she may stamp with her back feet and injure or kill the newborn rabbits.

Occasionally, a doe fails to produce milk. Then, the young will starve within 2 or 3 days unless the condition is noted and the young are transferred to foster mothers. Keep a close check on newborn litters for several days after birth to make sure they are being fed and cared for properly. If their bellies are rounded, they are eating regularly.

Expect occasional cannibalism

Does sometimes eat their young. This may result from a ration inadequate in either quantity or quality or from the nervousness of a doe disturbed after kindling. The doe may also be of a strain that exhibits poor maternal instincts.

Does usually do not kill and eat healthy young, but limit their cannibalism to young born dead, or those that are injured and have died. Proper feeding and handling during pregnancy will do more than anything else to prevent this tendency. Give another chance to a valuable doe that destroys her first litter; if she does it again, dispose of her.

What to expect from the litter

Newborn rabbits are small and helpless. By the 14th day, they have their eyes open and begin to hop around. Inspect their eyes at this time. Sometimes there is an infection or bits of bedding may be in the eye. Squeeze the eye gently and put in some eye ointment or cleaning solution, if necessary.

Your rabbits should remain in the nest for about three weeks. Occasionally one will show up earlier. Perhaps the bunny has been pushed

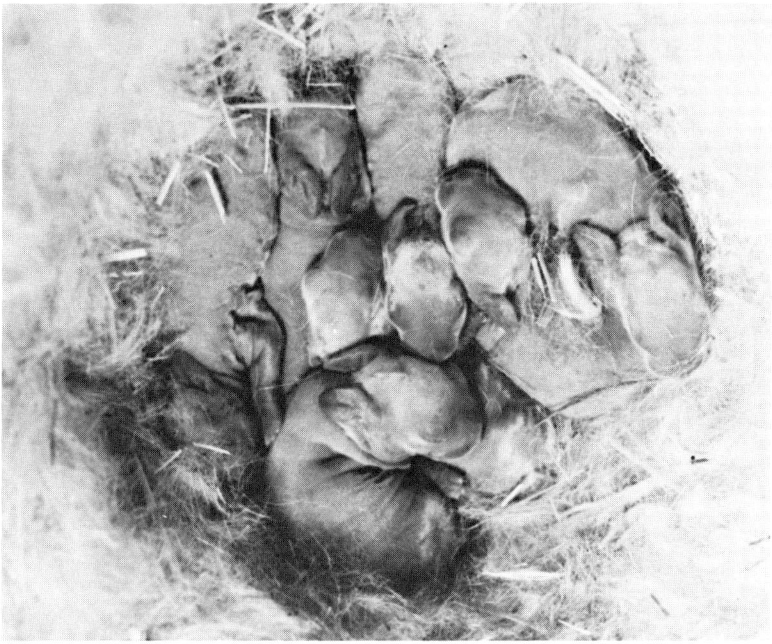

A litter of new-born bunnies in the nest.

out or has hung on to a teat and been carried out. Return him to the nest.

In four weeks the rabbits will be out and then they begin to share the doe's food. You will need to provide her with extra helpings. Many producers creep-feed the young rabbits with a special ration. Remove the nest box at this point to give the young rabbits more room.

After six weeks, the litter eats substantial amounts of feed and must have a ration that helps them become ready for the market at eight weeks.

Weaning the litter

Under most management programs, the young are weaned at 8 weeks of age. At this point, young meat rabbits should average 4 pounds in weight and be ready for market. Some commercial producers leave the young with the doe for 9 to 10 weeks to get a 4½ to 5½ pound fryer. Small litters (fewer than five young) can be weaned at an earlier age and the doe rebred. Also, under accelerated breeding programs, where does are bred less than 35 days following kindling, wean the young at 5, 6, or 7 weeks of age.

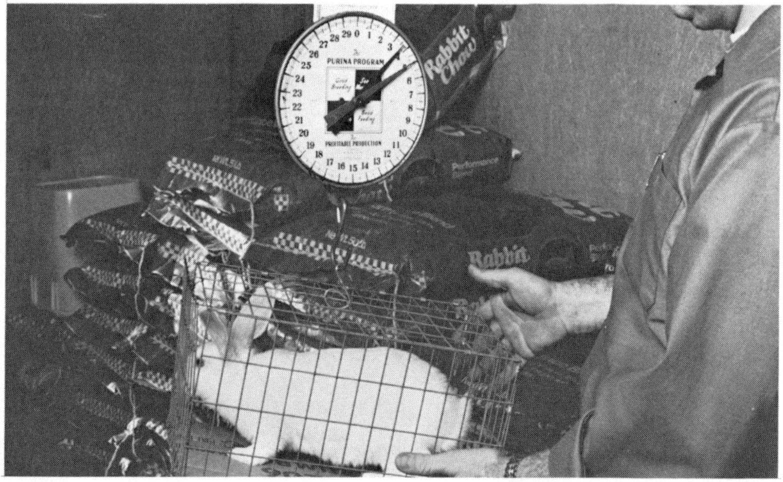

Plan to weigh each rabbit at weaning time

Check up on performance

Weaning time is when you should prepare a "report card" on the performance of the doe and the sire. For the commercial producer, the name of the game is pounds of rabbit produced. To improve, you'll want to save rabbits from the litters that have been the best doers.

The way to do this is to make it a standard practice to weigh each litter on a specific day—say at 56 days of age. You'll then have a standard for comparison of litters and their parents.

All of this calls for records. These take some time, but are absolutely essential, if you are going to make progress. If you are just starting out, you'll want to save the best does to increase your herd. Choose those from large litters which weigh the most at 56 days. Sure, you'll want to look at confirmation and over-all strength, but the ability to kindle big litters of fast growing rabbits is what you are looking for.

Litter weights also are a check-up on the manager (you). Have you done a good job of feeding, of providing water, and of giving other good care? Both litter size and weaning weight are a reflection on your management, all the way from selection of breeding stock through feeding and the other aspects of management.

Records you should keep

A hutch card is used by most breeders to keep track of the litters. A sample is shown here. Some feed manufacturers provide a similar card for their customers. It is a place to record the breeding date and dates of palpation, number kindled and number raised.

Set a standard for each doe in your rabbitry. We recommend that you expect each doe to raise 32 pounds of rabbit per litter. And her record for the year should be at least 140 pounds—and 150 pounds is better. If she can't do the job, mark her for culling. Many producers give a doe a second chance—but no more.

HUTCH RECORD CARD

Hutch No. ________ Ear No. ________ Sire ________

Record Page No. ________ Born ________ Dam ________

No. Nipples ________

Judge Milk Production by litter wt. at 21 days
Judge Meat Production by litter wt. at 56 days

Date Bred	Buck	Date Tested	Date Kindled	No. in Litter	No. Left		Av. Wt.		Remarks
					21 days	56 days	21 days	56 days	

BUCK RECORD CARD

Hutch No. ________ Record Page No. ________ Ear. No. ________

Born ________ Weight at 56 days ________ Remarks ________

Doe Bred	Hutch No.	Date Bred	Tested		Young Kindled	Sex		Wt. at 56 days
			OK	Missed		Bucks	Does	

Record performance information on hutch cards

Identify your rabbits

You must immediately identify young rabbits you select for breeding stock. A tattoo in the ear is the generally accepted way to do this.

A tattoo is permanent and will not disfigure the ears. Ear tags and clips are not satisfactory because they tear out and disfigure the ear. Identification is then lost.

Use letter and numbers of your own choice, but keep them to as few as possible. Consider how your subsequent numbers may be influenced by the first numbers you use.

Check to see that desired numbers and letters are the way you want them by testing on a piece of paper. Press them into *left ear* with tattoo punch. The left ear is used because the right ear is used for ARBA numbers. Then rub ink into holes made in the ear. Remove extra ink from the ear with vegetable oil. Wipe with alcohol to prevent infection. A felt-tip pen may be used for *very temporary* ear markings.

You'll also want to keep a record for bucks. It quickly becomes apparent which bucks have the ability to sire the big, profitable litters. Remember, the buck provides half of the characteristics of the young. And since he is used to service many does, his abilities are far more important.

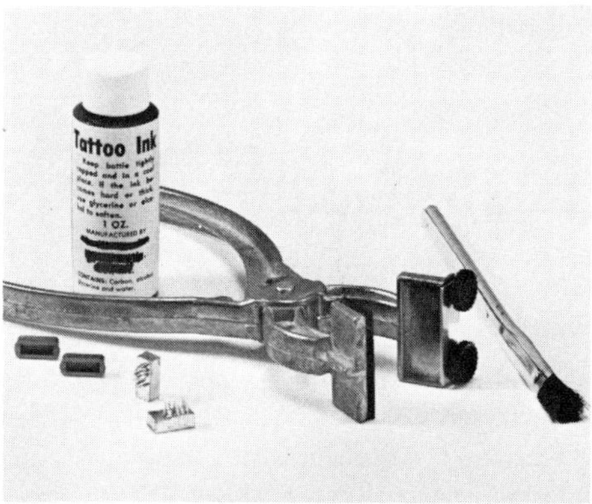

Use tattoo punch to put a permanent ID in the left ear.

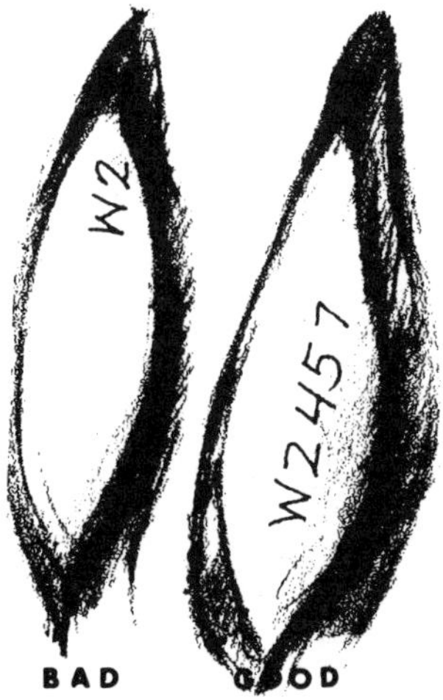

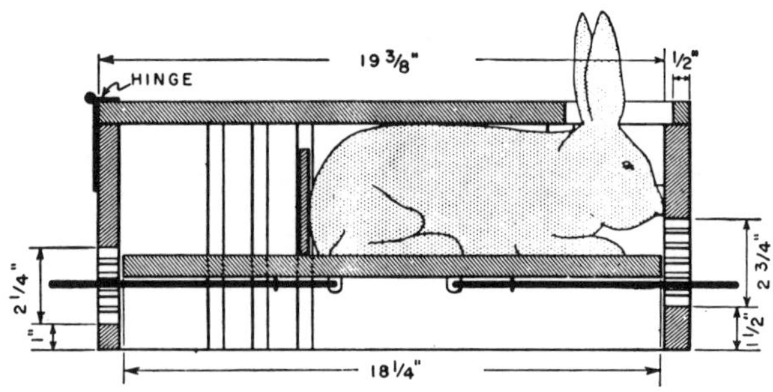

Build a tattoo box

You'll want to identify your rabbits as your business grows. In a litter of puppies, each one tends to be an individual probably with different markings. But try to identify a litter of all white rabbits!

The usual plan is to tatoo a number into the ear. A box that holds the rabbit close while letting his ears stickout makes the job easier.

This is a vertical section of a box for restraining a rabbit for tatooing. The spring-type holders tacked to the lower side of a movable floor compress the rabbit toward the top of the box. A movable cross-partition holds the rabbit toward the front. Blocks of wood on each side hold the rabbit's head in the center of the hold at the top. One person can do the job alone. It is almost impossible to hold a squirming rabbit long enough to get the tatoo just right.

Remember, rabbits may be precocious

After you select breeding stock, separate the young bucks and does. Anytime you leave the sexes together longer than eight weeks, you can expect some surprise pregnancies. Those little bucks are able to find willing does at an early age. Remember, gestation is only 30 days. There have been cases of 3-month old mothers...and that's just too young.

Keep a doe as long as she produces

When you get a high-producing doe that mothers her young well, keep her around. Your hutch record cards will show when she begins to drop off in productivity. Older does tend to milk better. They are ideal as foster mothers for some newborns from litters that are too large.

The average doe will have a productive life of about three years. This means that you will have to replace just over one-third of your brood stock every year.

The average useful life of a buck is four to five years. However, an older buck that goes through a spell of extreme heat in the summer may never be fertile again. Older bucks usually are poorer breeders in summer.

Cold and hot weather management

Extreme heat or cold is hard on rabbits, but they can withstand cold much better than heat. In cold weather, you may wish to use plastic or canvas siding to protect animals from drafts. The thinnest fur is on the belly, which is directly on the wire when the rabbit lies down. Place a board, gunny sack, or small square of carpet on the wire for the rabbit to lie on. Make certain you have a constant supply of water. Your rabbit cannot drink if your crock or water system is frozen. Put extra bedding in the nest boxes during cold weather; use a half-covered nest box.

Summer heat can cause high death losses unless you protect the rabbits. Shade for the hutches is extremely important. Make sure air can circulate freely through the rabbitry. A fan helps, but do not aim it directly at the rabbits. Sometimes it helps to thoroughly sprinkle the roof or even douse the wood inside wooden hutches. Sprinkle often during the day and early evening. Lay wet gunny sacks on the floor for the rabbits to lie on. Keep the sacks wet. If you have just a few rabbits, freeze water in milk cartons and place them on the floor of each hutch so the rabbits can lie next to them. Be sure that the rabbits have plenty of clean, fresh water available at all times. Most rabbits will not eat while it is hot, but they will drink frequently.

A rabbit does not perspire; the ears serve as its cooling system. If you cool the ears, you will cool the entire rabbit. If a rabbit is suffering from heat prostration or sun stroke, run cold water over its head and ears. Keep water out of the nostrils. Putting a block of ice on the head and ears and one under the body sometimes works well. You can also wrap a rabbit in a wet gunny sack, including the ears and head, but make sure the rabbit can breathe freely.

If a doe has a litter in hot weather, remove most of the fur from the nest box and save it in a closed bag or container so that you can put it back in at night. On hot days, a wire basket may help keep a young litter cool. Wire nest boxes work best in the hottest parts of the country.

Make a management schedule and stick to it

Rabbits need attention every day. Develop a routine and stick to it. For example, most rabbit producers feed in the evening. Rabbits don't eat much during the day so evening feeding puts fresh feed in front of them.

Waterers need to be checked every day. If you use crocks, clean and sanitize them on a weekly basis.

You'll need to keep a breeding "to do" sheet as your numbers increase. Remember, the timing of breeding and putting in the nest box is very important. It is something you can't keep in your head.

Cleanliness is one of the most important requirements for healthy rabbits. Thoroughly clean the feeders and waterers once a week. Scrub crocks, if you use them, with a strong liquid bleaching product. Clean and dry feed containers before putting feed in them. Regularly disinfect feeders and waterers with an effective, non-irritating sanitizer.

At least once a week, clean out fur that accumulates under and around the cages. Do it oftener in hot weather. A propane torch is effective for cleaning and disinfecting the cages.

Clean out the manure under the hutches or cages on a regular basis. Move it to an outside worm bed and work it immediately into the surface of the bed. Worms can be successfully raised under the cages. If you do this, water the worm beds regularly and turn the manure at

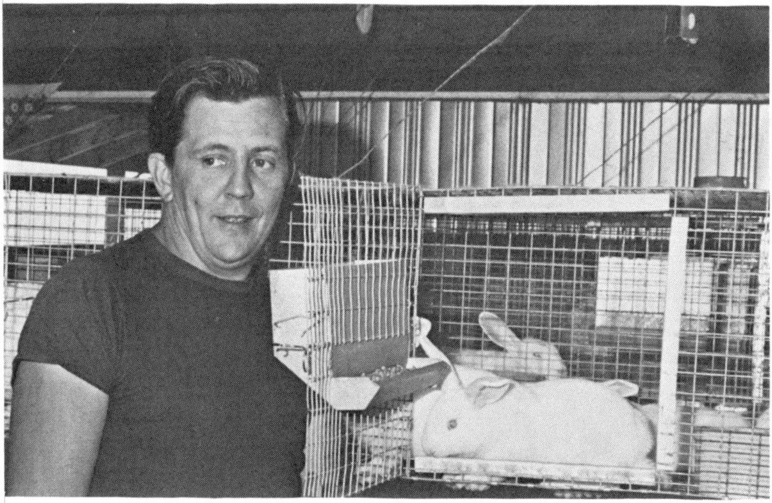

Inspect your rabbits regularly; keep hutches clean

least once a week. Keep aisles and work area clean. Avoid creating dust in the rabbitry.

Nest boxes must be thoroughly cleaned and free from all fur after each use. Wash and allow them to dry thoroughly in the sun. As you care for your animals each day, check them for any abnormal appearance to help detect illnesses.

These bunnies need a good ration to reach 4 pounds in 8 weeks

Chapter 6

Feeding for health and fast growth

To the novice, a rabbit is an animal that nibbles green plants around the edge of the lawn and steals lettuce and other goodies from the garden. Actually, that's about the way a wild rabbit gets along. But if you are going to have a profitable domestic rabbit, producing 40 fast-gaining bunnies each year, it requires some high-powered feeds to get the job done.

Virtually all commercial rabbits today are raised on a complete feed compressed into pellets. This contains a mixture of carbohydrates, vitamins, protein and other nutrients that keep rabbits gaining weight and producing the way you want them to perform.

The pellets are clean, easy to handle and you almost can be sure that your rabbits are getting the nutrients they need. Several different mixes are available to fit the stage in the life cycle of the rabbit.

Basically, there are three systems of feeding rabbits. The first and most popular is the all-pellet program. The second is pellets supplemented with green, leafy hay, garden vegetables and other plants. The idea is to cut down on the amount of purchased feed, while providing variety in the rabbit's diet.

The third is the old-fashioned way where the producer mixes his own grains, incorporates protein and supplements with alfalfa or some other greens. The homesteader who goes into rabbits to cut-down on grocery bills, probably won't want to turn around and spend money for feed. Others who farm and grow their own grain, look on home mixing as a money-saving proposition.

Start out with the pellets

My recommendation to anyone starting out with rabbits is to go with the pellets. You have enough management decisions, as you learn how to handle your rabbits, without worrying about mixing feed. Provide your rabbits with the recommended amounts of pelleted feed and they'll get the proper nutrients. Later, when you know more about the business, try adding some hay or other greens to the diet. Supplement the pellets with oats, corn, or other grains. But at first, play it safe.

Your feeding strategy

There are two major goals in planning a feeding program. The first is to keep breeding does and bucks from becoming too fat. This interferes with fertility and with long life. This means restrictive feeding for these animals.

The second goal is to give lactating does and their young all they can eat. The fast growth required to take those bunnies to four pounds in just eight weeks must be powered with a high-protein ration. Give mother and young the best ration you can find in this critical growth period.

Recommended amounts of feed

Here are some recommendations for feeding rabbits in the weight class of the New Zealand or Californian in a meat-producing rabbitry: They are based on using a pellet with 14% protein and less than 20% crude fiber.

Does at kindling: Feed two to four ounces the first day after the litter is born to prevent caked udders. Gradually increase at the rate of 1 ounce per day until you feed free choice within 4 to 5 days.

Lactating does and bunnies: Feed free choice as much as they will eat in 24 hours. Don't let stale feed accumulate.

Does at weaning: Limit feeding to 4 to 6 ounces per day. Cut back if the doe seems to be getting fat.

Gestating does: Let these non-lactating does have 4 to 6 ounces per day. Again, adjust to keep weight in line.

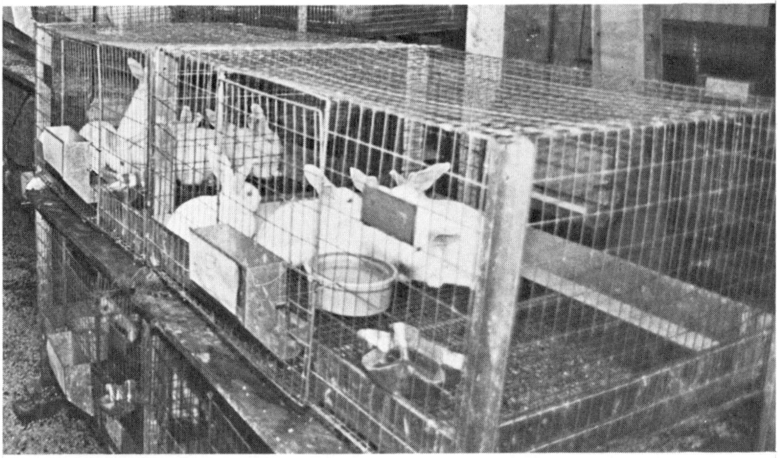

Cages with pellet self-feeders installed for easy filling from the outside aisle.

Weaned bunnies: You still want fast gains for the fryers. Let them have all they will eat free choice.

Dry does, bucks: Hand feed 4 to 6 ounces per day and adjust to maintain weight.

An extra shot for the doe

To improve milking ability, many breeders feed a little Calf Manna. This is a Carnation brand, commonly used as a milk replacer for calves. Feed one tablespoon per day, in addition to regular rations, at least a week before kindling. Continue until the bunnies come out of the nest and begin eating at the feeders. This extra ration helps the doe provide more milk in the critical first weeks when she is the sole source of food for the bunnies.

What goes into pellets

Pellets contain the leafy roughage, grain, protein, vitamins and other nutrients that rabbits need for fast growth. Usually ground, high-quality alfalfa hay makes up 40% to 60% of the pellet. The ingredients are ground, mixed and then compressed into a pellet about 3/16-inch in diameter and 1/2-inch long. These stick together quite well and do little of the flaking that causes waste.

Even young rabbits can handle pellets. And because they have been so carefully blended and researched, they can be the only ration fed.

Two pellet systems

Two major feed manufacturers have done substantial research with rabbit feeds. We are going to tell you about their recommendations in this chapter. We don't want to leave the impression that they necessarily have the best rabbit pellets. Many local mills can do the same quality job. But, since they are the major national brands and have representative products, their recommended systems are reported on these pages.

Purina Chows

The Ralston Purina Company, Checkerboard Square, St. Louis, MO, 63188, has maintained a rabbit research center since 1926. More than 150,000 rabbits have been fed out on the company's research farm.

Purina Rabbit Chow Complete Blend has at least 14% protein, 2% fat and not more than 20% crude fiber. This is recommended for all breeds of rabbits, large and small, in a normal breeding program up to 5 litters per year.

Purina Rabbit Chow Performance Blend has at least 17% protein, 2% fat and not more than 18% crude fiber. This is a blend for commercial producers who are striving for five to eight litters per year.

Carnation Rations

Carnation Company, Milling Division, 1700 Potter Ave., Kansas City, MO, 64126, produces four rabbit pellets plus Calf Manna.

Rabbit Family Ration is the basic adult feed for all rabbits. It is produced to be fed free choice to adult rabbits and their young.

Rabbit-Glo Creep Feed is made to be used in creep feeding young rabbits. It is a supplement to milk and the regular ration. This system takes nursing strain off the doe.

Rabbit Breeder Paks provide a developing and maintenance feed for pregnant does, junior does and breeding bucks.

Rapid Ade is a medicated feed that is recommended for periodic feeding during times of stress or disease outbreak. It is helpful in controlling coccidiosis.

These are just two examples of the way major companies are producing complete rations that make it easy to supply your herd with all of the nutrients they need.

How about medicated feeds?

You can select medicated feeds including a type of sulfa drug. They are helpful in controlling disease. Many feel the same job can be done more effectively through drinking water. The theory is that a sick rabbit may not eat but will drink. If you use medicated feed, shift back to regular ration after seven days. Healthy rabbits do better on regular feed. There are bacteria in the stomach that are needed to aid in digesting the food properly. A constant diet of medicated feed destroys these helpful "bugs."

Rabbits practice Coprophagy

Rabbits have a natural habit that may bother you a bit at first. Don't worry—it's all part of the rabbit's digestive system. They re-ingest part of their feces. At night, or early morning, they produce a soft stool that is eaten. You may never see this happen because they usually do this when unobserved. But if you do see this, don't think there's anything wrong.

Rabbits re-ingest only the soft matter that has passed through the digestive tract. Investigators have called this trait "pseudo-rumination," from the characteristic of ruminants (cows, sheep, and others) of chewing the cud, which is food regurgitated and chewed again. Most rabbit breeders are unaware of this practice. Some who have observed it believe it indicates a nutritional deficiency. It is, however, normal in rabbits and many actually enhance the nutritive value of the feed by virtue of a second passage through the digestive tract.

Watch your feed efficiency

Feed efficiency is a measure of the pounds of feed it takes you to produce a pound of rabbit. If a rabbit eats 16 pounds of feed to put on four pounds of meat the feed efficiency is said to be 4:1.

Broiler producers are the champion feed converters. They now can produce a pound of chicken with just two pounds of feed. Hog producers require around four pounds and beef producers eight pounds. The goal for rabbit producers is to be under four pounds or 4:1.

If your rabbit feed costs you 10¢ per pound, 16 pounds to produce a four pound rabbit will cost $1.60. If you sell that rabbit at 60¢ per pound ($2.40), your feed cost will be two-thirds of the total cost of production. This is why feed efficiency is so important.

It is difficult to measure feed efficiency for any single litter. That would involve weighing and recording feed every day. Do keep records of feed consumption by does and litters on a monthly basis. You then can figure your average efficiency over total pounds produced.

The best way to improve feed efficiency is to select the heaviest rabbits for breeding stock. Do this at weaning time. The fast gainers will have the genetic ability to transmit better feed efficiency. Over a period of time, you may be able to reach 3:1 or better.

A second way to improve is to avoid feed waste wherever you can.

The third thing is to keep a sharp eye on the ration. One brand or mix may have more grow-power but still not add to the cost of production.

What you should put in your own ration

We have suggested that the new breeder or the commercial breeder stick to commercial pellets. They are a sure way to provide the necessary nutrients. Pellets also save work for the commercial breeder. Can you imagine filling 200 individual feeders with a couple of handfuls of alfalfa hay every day?

However, you may feel that commercial feeds are too expensive. Or perhaps you like the idea of feeding natural feeds you put together by yourself. Here's how you should go about working out a ration formula with a good balance of nutrients.

Nutritive requirements of the rabbit

Protein: It has been established that the rabbit, like other rapidly growing animals, requires the proper quality and quantity of protein. This protein may all be of vegetable origin provided care is used in blending the various protein sources so as to obtain adequate amounts of the essential amino acids.

Protein is the muscle-building nutrient in the feed which is broken up into amino acids during its manufacture or digestion. There are several amino acids which in turn are building blocks for body proteins. Proteins are also referred to as nitrogenous nutrients which simply means that proteins contain nitrogen. Plants such as clover, alfalfa, and soybeans are rich in protein. These plants are called legumes.

Fat and Carbohydrates: These nutrients are required in the animal to supply the energy to live, grow and reproduce. Give rabbits grain, fat, or vegetable oil in winter to keep warm and two weeks before market to finish the meat with fat.

Carbohydrates are sometimes referred to as nitrogen-free extract (NFE). This group of nutrients contain hydrogen, oxygen, and carbon. Carbohydrates include foods such as sugars, starches, and cellulose. These are also called energy nutrients because they tend to give the animal energy. As the animal uses these foods it changes the carbohydrate to water (hydrogen and oxygen) and to carbon dioxide (carbon and oxygen) which it exhales from the lungs. Fiber is made up of cellulose which is a carbohydrate that is almost indigestible.

Fats are somewhat similar to carbohydrates but give more energy to the animal per pound of nutrient than carbohydrates; i.e., there is more available energy in a pound of fat than in a pound of carbohydrate. One percent added fat is good for rabbits in winter to help them manufacture heat. Recent discoveries indicate that rabbits prefer and use vegetable oil or fat better than animal fat in their diet.

Fiber: The rabbit is one of several species of animals which requires a considerable amount of fiber or roughage in its ration for a balanced diet. Fiber is usually furnished in the form of cured hay, green feeds, root crops, or oats. Fiber keeps the rabbit's system in order so that the manure stays in pellet form. Fiber will also prevent impaction (constipation) and indigestion, providing other nutrients are fed in proper proportions.

Minerals and vitamins: Little is known at the present time concerning the specific mineral and vitamin requirements of the rabbit. A deficiency of either minerals or vitamins will not occur when rabbits are supplied with a wide variety of feeds that include: 1) two or more grains; 2) a plant protein supplement; 3) a good quality legume hay; 4) green feed or root crop; and 5) common table salt. In order to supply the maximum amounts of vitamins, minerals, and protein, it is important that the legume hay be green and leafy when harvested. Vegetables should be fresh and sound.

Minerals are inorganic elements, such as calcium and phosphorus. Minerals are found in adequate amounts for the rabbit in good legume hay, such as alfalfa. However, the mineral sodium chloride or common table salt must be fed to the animal separately. Usually a block of salt is placed in the hutch. This is needed by all rabbits. It may be provided in the complete pelleted feed but should be fed separately when hay or roughage is fed.

Vitamins are chemical substances which are needed by the animal in very small quantities. But these small quantities are essential for the animal to properly absorb and use the nutrients in its body.

These four nutrients are best:

Roughage

- Chopped alfalfa hay—whatever amount the rabbits will clean up. (May substitute clover or other legume hay.)
- Green feed—about 1/10 pound daily, per doe, and fed only in racks above the floor where it cannot be contaminated. Alfalfa, grass, carrots, or other wholesome greens will serve. (May supply 60 percent of ration when hand-fed, or up to 30 percent when concentrates are fed in self-feeder.)

Grains and Protein Supplements

- These should make up about 40 percent of the total ration if hand-fed in self-feeders.
- Cereal grains, four parts, by weight, to include two or more of the following: whole barley, whole wheat, and whole milo. We suggest feeding wheat when old crop milo is not palatable. At other times, select whichever grain is the most economical. For herd bucks, dry does, developing does, and bucks, the concentrates may include two parts whole grain and one part protein supplement. Allow about 2½ ounces of concentrate mixture daily for animals of medium weight.
- Protein supplement, two parts by weight, selected from the following: soybean meal, linseed meal, peanut meal, or sesame meal. We suggest either pellet or pea-sized oil-cake form; pellets may be 3/16 inch in diameter and 1/8 inch in length. If necessary to feed in meal form, half the grains should be rolled to prevent the meal from settling out, and the mixture should be dampened just before feeding.
- Salt. Added as sifted-salt to the ground part of the ration or as salt spools, allowing 0.16 percent of the concentrates by weight. This would be 2½ ounces of salt in 100 pounds of feed.

Whole feed as compared to ground feed

Tests show that rabbits make as good gains on whole barley as they do on ground barley. Other grains were included in a consumption test and it was found that except for corn, all grains were consumed as well in the whole state as when ground. When soft corn is fed, the whole kernel is consumed. The hard or flint corn kernel must be cracked or rolled because if left whole, the animal will eat only the germ of the kernel.

Variety and palatability

Rabbits have certain feed preferences. Grains are listed here in order of preference—oats, wheat, milo, barley, and corn. Feeding too much barley, milo, or corn may result in pelts with coarse, stiff, dull hair.

Pelleted protein supplements are preferred in descending order: peanut meal, soybean meal, sesame meal, linseed meal, cottonseed meal, or mill by-products, which in the form of mash or pellet feed, give considerable variety.

Fresh green plants make good feed

Rabbits relish green plants. Here is a listing of plants that can be fed to rabbits:

- Bermuda Grass
- Bur Clover
- Chicory
- Clovers (red clover, white clover, alsike clover, white sweet clover)
- Dandelion
- Filaree, alfilaria, or stork's bill
- Foxtail or wild barley before heading
- Johnsongrass, first cutting
- Lettuce, wild or prickley
- Malva or mallow, cheeseweed
- Morning-glory, wild
- Mustard, wild black
- Oats, wild—when green
- Plantains, common and English, or ribwort
- Thistle, Napa and yellow star

Plants that cannot be fed to rabbits:

- Burdock
- Chickweed
- Sour clover
- Fireweed
- Goldenrod
- Horehound
- Lamsquarter and similar species
- Lupine
- Milkweed
- Miner's lettuce
- Poppies
- Tarweed
- Turkey mullein

Supplementing pellet rations

You can cut down on the amount of pelleted feed you have to buy by supplementing with leafy green hay or other home-grown feeds. Dr. Peter Cheeke, Oregon State University Researcher, reports that he has done some work on feeding greens to rabbits. These include clover, lettuce, celery and grass fed free choice to weanling rabbits.

Dr. Cheeke says that it is possible to reduce intake of pellets by 50% with no reduction in growth rates. This means that if a rabbit is eating 100 grams of pellets per day, his ration can be cut to 50 grams of pellets with greens free choice, without sacrificing performance.

Remember that greens are high in water content. A large volume must be consumed to get enough nutrients. Watch the quality of greens carefully. They can't do the job alone. Perhaps an adult rabbit can get along on a maintenance situation but greens will not support growth or lactation.

For home rabbit production, Dr. Cheeke says a ration of hay and salt free choice, with rolled oats, barley or corn, fed in limited amounts, should be satisfactory. A bit of testing can be done to find the minimum level of rolled grain that will support the desired level of production. Such a diet would be a bit low in protein; a small amount of soybean meal could be added to the grain.

The hay should be of good quality, and preferably contain clover or alfalfa. The legume will supply protein, calcium and trace elements, while the grain will provide calories, some protein and phosphorus. Fat-soluble vitamins (A, D, E, K) will be provided by the hay and grain, as will the B vitamins, which are also synthesized in the rabbit's intestinal tract.

It is possible (and desirable) to raise rabbits on a backyard scale using home-grown feeds. However, for commercial production, the high labor cost of feeding hay and greens, and the reduced level of animal performance generally, make pelleted feeds a better bet.

Water is the cheapest feed

Rabbits require large amounts of water to make good use of feed. If rabbits do not have a supply of fresh, clean water, they will not eat as they should. Their bodies require a lot of water in the feed utilization process. If you notice rabbits are not eating, the first thing to check is the water supply.

Don't share the feed with pests

Rabbit feed is expensive. Make sure you reserve it for rabbits alone. Unprotected feed attracts rats and mice, insects and birds. All of them can create problems for your rabbits, as well as costing you money.

If you are a small producer, store your feed in a large metal garbage can. Don't get too much on hand at one time because it can become stale or moldy. Both conditions are bad in a rabbitry.

For larger numbers, provide clean, dry, well-ventilated storage. Keeping feed as cool as possible, helps preserve quality.

A good litter of rabbits ready for market

Chapter 7

How to find a profitable market

The day you bring your first rabbits home, you have opened a business with many opportunities and challenges. You actually are operating a food factory and face the same management decisions as anyone else going into food production and marketing.

The key to success in business is the ability to sell the product at a profit. No matter how good you are at producing fryer rabbits, you still must have a plan to sell the meat. Your factory produces by-products too, including breeding stock, fur, manure and even rabbit's feet. Each must be operated as a profit center, building total income for your enterprise.

Marketing is harder for most rabbit breeders than production. You probably go into rabbits because you like them and want the satisfaction of raising them. The idea of beating on doors to sell the meat is something you'd rather not do. Chances are, however, that's what you must do to move your product.

There's an old saying, "If you build a better mousetrap, the world will beat a path to your door." Don't believe it! What happens in today's world is that the man who builds a better mousetrap has to go down the paths, beating on doors. And that's after he sharpens his price to meet the competition.

Plan your sales program

A sound sales plan is the first step in rabbit production. Before you buy the first rabbit, make sure you have a good idea of where you can sell them. If you begin with our suggested three does and a buck starter unit, within about four months you'll have 25 rabbits ready for the frying pan. Sure, some of them can feed the family and friends. But right then you'll need to have a marketing plan for the remainder. Three months later, you'll have another 25 rabbits and so on all through the year.

Some part of every week must be spent in marketing, regardless of the size of your operation. You can't afford to hire help, so you (or your spouse) will have to divide up the marketing functions—sales manager, sales promotion manager, advertising manager and public relations manager. In this chapter, we want to show how all of these important jobs can be done.

The ways to market rabbit meat

Let's examine some of the ways you can sell meat rabbits. You won't be able to crack the larger markets at first. Plan to build your own territory and list of customers. Later, it may be possible to find one of the larger processing markets. We'll list them first, then describe them in more detail.

Major processors: These large operations sell rabbit meat all over the country. They have their own labels and they have access to supermarkets. They buy live rabbits. Some pick up, others accept live rabbits at the door. It's a good clean operation that leaves you free to concentrate on production. There's no messy slaughter and you do a minimum amount of selling. The bad news...it's the poorest paying market for your rabbits. The other disadvantage is that processors have a bad habit of coming and going. Hundreds of rabbit producers have built up an operation to supply a processor only to wake up one morning and find their market is out of business.

The local or regional processor: There are substantial numbers of these processors who build a trade around one urban area or a portion of a state. They solicit a grocery store trade and keep it supplied, usually within the area that can be served by a truck or two. They build slowly to match the demand and may be more reliable than a larger operator. Again, the price to the producer is low because he gets none of the middleman dollars.

Do your own processing: This involves setting up a unit that can pass local or state inspection. You process your own rabbits and build a route to supply retail outlets. You do the selling and receive the mark-up for both processing and marketing.

Sell live rabbits and butcher them: Many states make it tough to process rabbits through regulation. Producers in this situation often sell the live rabbit, then process it for the buyer for a fee.

How much can you expect to receive?

In 1982, the major processors were buying rabbits for about 60¢ per pound, live. They process and sell to the supermarket for $1.60 per pound. The average rabbit dresses out at about 50%. This means that the processor who paid $2.40 for your rabbit sells it for $3.20 to $3.50. He's lucky to have $1.00 margin for his processing and marketing work. Of course, he does have the pelt to help offset his costs.

Pel-Freez frozen rabbit is sold in a giant supermarket in surburban Milwaukee, where we shop. Frozen rabbit is in the case with crab, shrimp and other top-line meats. Early in 1982, a 2.25 lb. package of frozen rabbit was selling for $2.43 per pound. That's $5.47 for a rabbit meal.

Across the aisle, in an open freezer case, broilers were selling for 89¢ per pound. A family could pick up a chicken dinner for $2.25.

This outlines the problem for the rabbit producer. Rabbit meat must be sold on its quality. It can't compete with chicken, pork or the less expensive cuts of beef. It is priced in the lower range of steak... and if most families have to choose, it will be steak at this point in time.

In effect, the widely distributed Pel-Freez frozen package sets the upper price limit for rabbit. Your goal should be to get as much of that $2.43 per lb. retail price for yourself as you possibly can.

How the major processors do it

Pel-Freez: This well-known processor is located at Rogers, Ark. and has been in the rabbit processing business for many years. The company is best known for its well established brand of frozen rabbit. The company processes two million pounds of rabbit per year and buys 16,000 to 18,000 live rabbits per week. Trucks radiate out from Rogers to pick up live rabbits on a regular basis.

The company formerly operated a plant at Anton, TX, but that has been closed. Other routes have been shortened to cut the cost of hauling. Close-in growers have been encouraged to expand. Around 500 growers supply Pel-Freez.

Pel-Freez buys only white rabbits. And the company maintains high standards for quality. It takes only healthy rabbits, in the right size

Pel-Freeze rabbit is in the frozen food cabinet in this giant supermarket. It's sold with the specialty meats.

Pel-Freez is a recognized national brand. Freezing helps balance ups and downs of supply.

and condition, for the level of product quality that has been established.

If you were a Pel-Freez producer, you would receive 58¢ per pound for live rabbits as a base price. In addition, there is a bonus for volume production. In 1982, it was as follows:

Pounds Sold	Bonus rate per pound
3,000 to 5,999 pounds	1/2¢
6,000 to 9,999 pounds	1¢
10,000 to 14,999 pounds	1-1/2¢
15,000 to 19,999 pounds	2¢
20,000 to 24,999 pounds	2-1/2¢
25,000 and over	3¢

Pel-Freez also has a program for Certified Grade A producers. This is based on location, size of the rabbitry and the amount of rabbits produced in a 13-week period. It pays 3¢ to 6¢ per pound.

The turn-over in producers is relatively infrequent. While some drop out each year, others come into production. There usually is a waiting list of producers who want to get into the plan.

Rogers, Ark., is in an area that has been a major broiler producing area. Many rabbit growers utilize remodeled broiler structures and have experience in broiler production. They are accustomed to high production on narrow margins.

Hillcrest Foods. This processor operates in Hill City, Kans. The organization has changed hands twice in recent years but continues to market large amounts of rabbit meat. Hillcrest can process 5,000 rabbits per day and this is comparable in size to Pel-Freez.

Rabbits are purchased from eight states including South Dakota, Nebraska, Iowa, Minnesota, Missouri, Kansas, Colorado and Oklahoma.

Size of the rabbitries runs all the way from five does to the giant Circle J Ranch at McCook, Nebr., which has 1,500 working does and 20,000 rabbits on the place at all times.

Rabbits are delivered by the owners directly to Hillcrest. Usually, one producer operates a truck route for others. When the rabbits are weighed in, the producer takes home a check.

Early in 1982, Hillcrest was paying a straight 60¢ per pound, live. Every producer, large or small, gets the same price for a rabbit of comparable quality.

Hillcrest freezes the rabbit meat for shipment, but encourages retailers to sell it as fresh. The rabbit is cut into eight pieces, put on a tray and covered with a plastic bag. A vacuum-pack process is used in which the air is replaced with CO_2 gas. The meat has a 28-day shelf life in the supermarket refrigerators.

The package is shipped out frozen. Stores put it in the meat counter, where it thaws overnight. Housewives then buy it as a fresh product.

Hillcrest is trying to build the market. Salesmen attend food shows and work hard at gaining acceptance in the supermarket. Recipes and other sales promotion materials are provided.

USDA regulations require processors to provide an inspector and pay him. Other meats have inspectors provided by the federal government. Because of this rule, only Pel-Freez and Hillcrest Foods have an inspector. This gives them broader opportunities to market the product in the United States and through export markets. Inspected meat can be sold to institutions and to the Armed Forces.

Regional markets

Grand Ronde Rabbitry, located in Grande Ronde, Ore., is a good example of a regional processor. This organization sells over much of

Oregon including Portland and Eugene. It processes from 300 to 500 rabbits per week and could handle as many as 1,000.

The owners work at developing grocery store trade for their rabbits. They feature a whole bodied rabbit displayed in a meat tray. A sausage market also is being developed.

Grande Ronde buys rabbits from about 25 producers in the general area. Their current price is 60¢ per pound delivered to the processor. Wholesale prices are in the \$1.50 to \$1.70 range. Retail market prices in the Northwest run from \$1.98 to \$2.69 per pound.

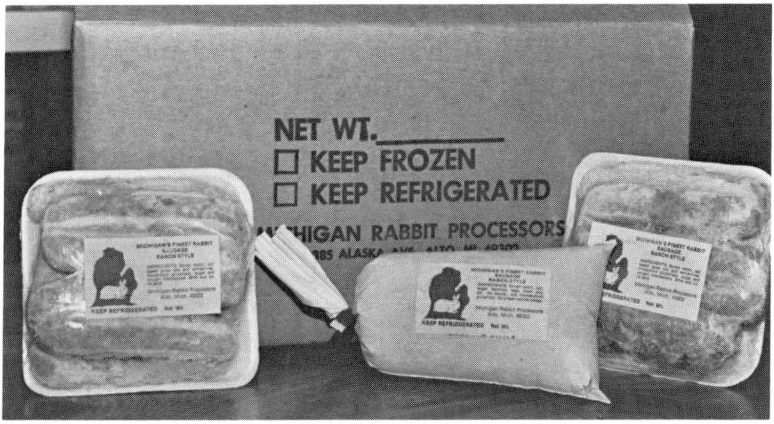

These packages were developed by Michigan Rabbit Processors to establish identity for its regional brand.

Michigan Rabbit Processors, Alto, Mich., bases its operation on rental of time in a large animal slaughter house. They use the facility two days a week when other livestock is not being processed. The freezer facility also is included in the deal.

Use of an existing facility cuts investment and avoids setting up to pass state regulations. Once a license is issued, it usually will cover rabbits, as well as other meat animals. Many rabbit producers make arrangements with local poultry processors to use their facility or to have the butchering done.

Michigan Processors markets rabbit meat through five grocery store chains. They also have developed some restaurant trade. Custom packages with a distinctive label have been developed for both fryers and rabbit sausage.

Around 25 producers provide the rabbits for this firm. The firm pays 58¢ per pound for white fryers to those who deliver occasionally and 60¢ to those who sell regularly. These are the prices for white rabbits. Colored rabbits are bought at a 7¢ discount.

All rabbits are delivered to the plant. Around 200 are processed on an average day. The firm stocks enough frozen rabbit to make up for fluctuations in supply. However, the market prefers fresh rabbit as much as possible or frozen that has been thawed overnight. Fresh rabbit sells at retail in the $2.30 per pound range. Rabbit sausage, made from older rabbits, is sold in bulk for about $2.00.

A local market

Ralph and Edna Wehr, Salem, Ohio. This family operation has developed a small, but continuing rabbit meat business in the Youngstown urban area. They raise a New Zealand cross bolstered with one-fourth Flemish to get more size. Fryers dress out at 4-1/2 pounds. They also buy rabbits from other producers to fill their orders.

The Wehr's got into processing when a projected local plant fell through. They put a processing set-up in the basement and it passed state inspection. They have developed a market in supermarkets and a couple of restaurants. They also supply two nursing homes with rabbit on a twice-a-month basis. The director of the home recognizes the nutritive value of rabbit for retirement people. This is a market that takes 30 rabbits with each order.

The Wehr's are paying 55¢ per pound for rabbits and wholesale a dressed rabbit for $1.65. They figure they get about $1 per rabbit for dressing.

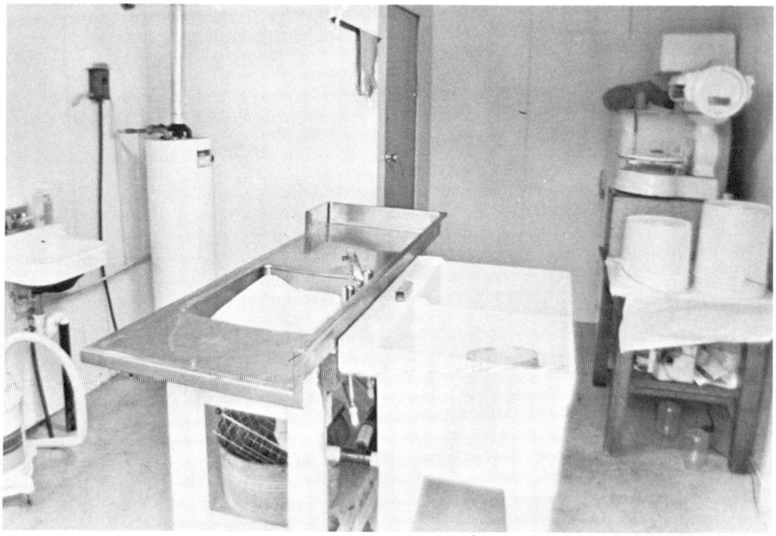

A small, well-arranged processing center

The Wehr's have developed their market through hard work and promotion. Ralph is a carpenter and when he is not busy he promotes the rabbits. The first thing he did was put up a professional looking sign on the front lawn, telling the world they had rabbits to sell. This brings at least a half-dozen customers each Saturday who buy anywhere from one to 10 rabbits. A local newspaper ran an article about the rabbitry. It subsequently led to a TV appearance. The Wehr's talk rabbit whenever they can and are growing toward sales of 2,000 per year.

The biggest problem is a year-round supply of meat to fill orders. In winter months, when demand is good, the supply of rabbits is too small. In summer, when demand is low, the big spring crop is ready for market.

This is the same problem faced by processors of all sizes. They need producers who are willing to work to solve the problems brought on by heat and cold, in an effort to keep a market that has been built up.

Wilde Hare Rabbitry, Sheridan, Wyo. Sawley Wilde is a small producer who doesn't feel that he wants to meet the stiff demands for a state processing license. He sells rabbits to friends, neighbors and others on an individual basis. At the request of his customers, he will butcher the rabbits, cut-up and package.

He finds that even when sales are made on a personal basis, people want an attractive package. Sawley packs rabbits in a meat tray he buys from a local butcher. The rabbit is garnished with parsley and covered with plastic wrap. He says you can't place enough emphasis on cleanliness, proper cutting and making the package attractive.

Take a look at all of the markets

Rabbit meat is the only big volume market for your herd. But for the small producer, it also pays to explore the alternatives. Sell breeding stock, sell pets and seek out laboratories and research centers. All of these return more dollars but they also require a higher level of management. The requirement for high-level performance runs all the way from selective mating through records and persistent salesmanship.

Breeding stock sales

It takes a reputation to get top dollars for your breeding stock. This can be built in the showring for the fancy breeds and to some degree for the meat rabbits. However, the main thing a buyer of breeding stock for meat looks for is production performance. You need the records to show that your rabbits were heavy-weights at weaning time, that they came from large litters and have a history of good health.

A few breeding rabbits go as low as $5 each, but $15 is a common price for a good individual with papers. Others with a blue ribbon background on the show circuit go at $35 to $50.

If you compare 60¢ per pound from a processor with $15 per head, breeding stock looks like the way to go. However, breeding stock is a limited market and not one that is easily tapped by a beginner. Only the pick of the litter should be sold—that leaves a lot of meat rabbits for you to sell. If there's a chance to sell breeding stock, the kind you are proud of and willing to back up, the money is good.

Cute, small-breed rabbits are most often wanted for pets.

The pet market

A rabbit breeder can sell a few rabbits for pets each year. Easter is the bonanza time for the pet market. Christmas is good but not in a class with Easter. Some breeders find they can sell a few year-round. The large meat breeds are not as popular for the pet market as some of the fancy types. Smaller rabbits do better than larger breeds. Color markings and nice fur help attract attention.

Pet market rabbits should be in the six to eight week range by holiday time. They are old enough to get along on their own at this age. It takes some pre-planning to get a stock of rabbits at just the right age for the Easter or Christmas markets.

Most people who shoot for this market, try to line-up pet stores in advance, for the bulk of their sales. This helps the grower decide how many to breed for the holiday trade. Others advertise pets and sell right at the rabbitry. Pets are sold directly in the $5 to $10 range wholesale. Pet stores double the price to retail customers. If you sell pets yourself, offer some pellets along with the rabbit. Your customers are likely to have more success with their new pet if the rabbit is not exposed to a sudden change in feed. Mark-up the feed bag enough to make some extra profit on the deal. You may even want to develop a continuing feed supply business.

A three-way profit plan

For some producers, more than one breed of rabbits fits the market in the area. Board Valley Rabbitry, Sparta, Tenn., has New Zealand Whites and Reds, Californians, Rex, Satins and Dutch. Here's how they fit the market:

The New Zealand Whites and Reds and Californians are sold for meat through newspaper ads and direct to meat markets. The Satins and Rex rabbits are sold for fur with some meat in addition. Dutch rabbits are favorites for pets. They are sold at flea markets and pet stores. More than one breed complicates breeding and record keeping. But it can spread out the market for production.

The laboratory market

Rabbits are much in demand for use in medical and biological research. As many as 700,000 are purchased for this purpose each year. New Zealand White rabbits are the breed most often ordered for research purposes.

On the average, you can get twice as much for a laboratory rabbit as you can for a meat rabbit. Special orders bring premium prices. Many people in the business feel they more than earn the premium. The quality of the rabbit required is nearly flawless. The rabbit must be in good health; with no ear mites and snow-white in color.

Researchers have a bad habit of ordering today and wanting the rabbits tomorrow. It is necessary to anticipate what those orders are going to be. The researcher expects his supplier to have what he wants, when he wants it. In many cases, the demand is fairly regular but a week's notice is about what you can expect.

The orders placed by biomedical research institutions can be categorized into several groups. The sex of the animals ordered is one group. The order may specify all males, all females, mixed, or either sex. Another grouping is based on age. The order may request six-hour-old rabbits, six-week-old rabbits, or six-month-old rabbits. A third grouping is weight. The order may specify four-pound to five-pound rabbits or 8.5-pound rabbits or 3-kg. rabbits.

Some orders may be filled by using rabbits which have not yet been weaned and are thus still with their mothers. But some orders are for six-month-old rabbits. Mature female virgin rabbits often are ordered for teratology. Hundreds of rabbits may be required at one time. To be able to fill orders of this type, large inventories must be maintained.

As you can see from these requirements, laboratory markets for rabbits demand a high degree of sophistication. Equipment and facilities are often inspected by buyers. It takes both size and investment to win in this game.

Since demand is erratic, the laboratory rabbit supplier must also have a regular meat market. He has cull animals to sell as well as those that just weren't right for a particular laboratory order...or those grown in anticipation of an order that never came.

The U.S. Department of Agriculture maintains a list of Registered Research Facilities that use rabbits and other animals. Write Senior Staff Veterinarians Office, USDA 703 Federal Building, 6505 Blecrest Road, Hyattsville, MD 20782. Animals for Research, A Directory of Sources is available from National Academy of Sciences, 2101 Constitution Ave., Washington, DC 20418 for $7.25.

Another source of information is American Association of Lab Animal Science, Suite 205, 210 N. Hammes, Joliet, Ill. 60435.

The fur rabbits

The fur of Rex rabbits is in great demand and bring premium prices. The Fur Growers Co-Op, P.O. Box 192, Eagle Creek, Ore., 97022, was organized to help Rex producers market furs. It has been able to get $12 to $13 for a Grade A pelt, $8 for a Grade B pelt and $4 for a Grade C pelt for producers. Pelts are graded according to fur quality, based on density and consistent color.

John and Janet Snapp, Morgan Hill, Calif., have 100 Rex breeding does and produce over 3,000 rabbits each year. They also have set up a tannery to process and market Rex fur for themselves and other producers. They report Rex furs selling at an average of $10 each. Chins, Castors, Whites, Browns and Blacks are in greatest demand.

The challenge is to have enough Rex fur of a specific color to interest a fur buyer. The people who manufacture coats, jackets or other garments want a uniform lot. While Rex production is growing, a shortage of breeding stock is limiting the number of furs available.

An example of the advertising that calls for careful study.

Rex rabbits are raised like other rabbits. However, when they are weaned, they should be kept in individual hutches until they are 16 to 24 weeks of age. The fur then has an opportunity to mature and reach the prime stage. This plan requires a substantial increase in the amount of housing, feeding and watering equipment. If you save six rabbits from a litter, that means six times as many hutches as in a meat operation.

Not all Rex rabbits have a marketable fur. These can be slaughtered for meat. The Snapps own a grocery store and they sell fur culls through the meat department. A Rex at 16 weeks of age still makes a tasty fryer. The Snapps figure that meat and fur sales combined bring them a profit of about $200 per year for each doe.

The dollar potential of Rex fur has brought a number of high-powered promoters into the field. The typical ad shown on this page has a headline, "Make $25,000 per year." The fine print says a minimum $1,250 investment is required. Be careful of these "buy-back" promotions. The chances of making big money in any type of rabbit production are quite small.

Sell wool from Angora rabbits

Angora rabbits are a special breed raised almost exclusively for their wool, although meat is a by-product. The rabbits are sheared much like a sheep every three months. Many growers raise Angoras because they like to spin the wool themselves. Others sell the wool at $5 to $7 per pound to spinners. Local craft shops can help you find buyers. Wool on Angoras grows to a length of 2-1/2 to 3-1/2 inches each 3 months, or approximately 1 inch per month. You can shear 14 to 15

Wool is shorn from Angora rabbits on a regular basis.

ounces of wool a year from a mature Angora that is not nursing young. This wool is valued for its softness, warmth, and strength. It is used in blends with other fibers in the manufacture of children's clothing, sport clothes, garment trimmings, and clothes for general wear. Used alone it is usually too light and fluffy, and blends create better tensile strength and durability.

Keep bucks and does in individual hutches. Keep woollers—does and castrated bucks maintained primarily for wool production—in groups or colonies to save labor. Castration of bucks that are to be reserved for wool production may reduce fighting in the herd, but wool production is not increased by castration.

To prevent infestation with internal parasites and to keep wool clean, install self-cleaning floors in the pens.

Wool should be harvested prior to breeding to prevent mauling, and soiling of the wool. Angoras are generally sheared or plucked every 10 or 11 weeks though some producers pluck their animals monthly and some at intervals beyond 3 months.

Promotion is a must

The No. 1 job of every rabbit producer is to promote his product. People in general simply are not aware of the value of rabbit meat, fur or by-products. You have a product to sell and it takes advertising and promotion to get the job done. Following are some of the ways to build interest in your rabbits.

Here's a sign that says "We have rabbits."

An attractive sign

The first step is to put up a sign that tells the world you are in the rabbit business. We suggest that you have it done by a professional. A hand painted sign frequently gives the wrong impression. Establish the name of your rabbitry in the minds of people in the community. If you process meat, list that service. Mention pets, furs, earthworms, fertilizer and anything else on your "billboard." If you are not on a heavily travelled road, consider putting a directional sign where it can be seen by a large number of people. We think Ralph Wehr's sign, shown here is an excellent example of a sign that says "we're in business."

List in the yellow pages

If someone is interested in rabbits, they are likely to "let their fingers do the walking" through the yellow pages. You can list your rabbitry in the classified directory for a relatively small amount each month. Again, list meat, processing, pets and anything else you have to sell.

A check in our metropolitan Milwaukee directory reveals there's not a single rabbit listing. The rabbit category says "see pet shops." Only a few of the pet shops mention rabbits. This urban area with more than a million people offers a market that is being ignored. Similar opportunities exist in other cities.

Newspaper advertising

Most rabbit breeders agree that a small classified ad, run regularly, pays off better than a large one run only occasionally. In smaller communities, you can run a two-line ad for $2 per week or less.

Lynn & Bill's Bunny Farm, Locust Grove, Okla., uses this ad in a near-by daily paper:

"Fresh rabbits, whole or cut up. $3.50 each. State Licensed. 386-2251."

It costs $21.50 per week and Lynn figures it pays well. The same ad in a near-by weekly costs $2.50.

Radio advertising

An announcement on radio can have strong pulling power. Richard Bombren uses radio to market meat from his Sleepy Hollow Rabbitry, Triadelphia, W. Va. An FM station that plays contemporary "background" music was selected. The station was rated second in audience in the area. Here's a sample of the first commercial:

> "Without a certain amount of knowledge of the various types of meat, many customers hesitate to spend precious food dollars on meat they are unfamiliar with. This is a disadvantage since it minimizes variety in menu planning.
>
> Pound for pound and dollar for dollar, rabbit is one of the best meats you can serve your family. Rabbit is higher in protein than chicken... it is low in cholestrol and has no fat. It's a delicious alternative to other meats. So next time you're considering what to serve your family for dinner, choose rabbit. They'll love it. This message is from Sleepy Hollow Rabbitry, serving the Wheeling area. Sleepy Hollow features only USDA inspected fresh meat. Our rabbits also are available at Busack's IGA in Belair and at Valley Cheese Co. in Central Wheeling. Call us at 547-1231. That's Sleepy Hollow Rabbitry—They know what good is..."

This is a 60-second commercial that was scheduled 3 times every day for a total of 30 per month. Cost was $160, but the two markets mentioned pay half of the cost. After the basic marketing story was established, the commercials were cut down to 30-seconds. Results: After the radio commercials went on the air, sales went up from 10 or 12 a week to nearly 100.

Use those bulletin boards

Every community has places where you can post "for sale" notices. You'll find them in grocery stores, sale barns, feed stores and other places. Make up a card that lists what you have to sell and pin it up in these places. Cost is very low and you'll reach a lot of people. Do a neat job of typing your card or have it printed. Be professional in every promotional activity.

Encourage people to taste rabbit

Roger Draper, Magna, Utah, likes to have a booth at celebrations, picnics and other church and civic outings. He provides samples of ready-to-eat rabbit to all who want to taste. It's a free-of-charge offering that helps overcome one of the big problems of the rabbit business —convincing people that rabbit meat is a taste treat.

This van complete with sign stops prospective buyers.

A mobile "store"

A truck or van, parked along a busy highway, offers another way to introduce and sell rabbit meat. Lynn Cox, Lynn & Bill's Rabbitry in Oklahoma, takes her Dodge van out on the main highway to Tulsa between 3 and 6 p.m. each day. Her "mobile store" is exposed to as many as 20,000 potential buyers each week. She has many repeat customers but more important, she is reaching new people every day. Sales run from 25 to 50 rabbits per day at $3.50 each.

The mobile store idea works well for the Easter pet market too. Park in a shopping center or near-by where Mom and the kids can see the rabbits on the way to buy their new Easter outfits. Put up plenty of signs.

Be prepared for the sale because the kids won't let Mom say "no." Have a carry-home box and a sack of feed ready to go along with the rabbits.

Show your rabbits

Many small producers develop very good markets for live rabbits by showing and selling at local fairs or swap meets. We're not talking about the very competitive shows for fanciers. Look on these lesser events as a means of getting your name and rabbits in the public eye. Have good signs for your display, along with business cards to give every prospect.

Prepare a simple brochure

You need something to hand-out to every visitor and every person you talk to at a show, fair or other outing. We suggest a single piece of paper folded in half to give you a 4-page piece that is more or less pocket sized. It can be typewritten but we think printing at your local newspaper or job shop is worth the money. Again, you are putting material in the hands of people for the first time. Create the professional image.

Here are sample pages from two simple flyers prepared by Seacord's Ed-Mar Rabbitry, Los Gatos, Calif. One brochure is for meat sales and the other for breeding stock. The brochures include information about the superior quality of rabbit meat and recipes for its preparation.

Business cards

Have business cards to leave behind when you call on stores, or other potential markets. A few hundred extra, printed at the same time, cost very little. Hand one out whenever you talk rabbit. It's a great way to remind people that you have the rabbits or rabbit meat they want.

We like this card, also from Edward Seacord. It tells everything a prospective customers needs to know about the products offered by Seacord.

Front

SEACORD'S ED-MAR RABBITRY
(A.R.B.A. REGISTERED RABBITRY NO. 2275)
17015 POLLARD ROAD
LOS GATOS, CA. 95030
PHONE (408) 378-8976

TOP QUALITY BREEDING STOCK

Californians

White New Zealands

Champagne d'Argents

Since 1945

"A BUSINESS, NOT A HOBBY"

"E D M A R"

T O P Q U A L I T Y - R A N C H R A I S E D

D R E S S E D R A B B I T S

F R Y E R S

R O A S T E R S

S T E W E R S

Back

DOMESTIC RABBIT MEAT
PROVIDES THE HIGH PROTEIN MEAT NEEDED

To date, there is no meat with a proven protein content that will equal Domestic Rabbit Meat. Domestic rabbit meat is a delicious, delightful tender white meat as distinguished from the "gamey" and dark meat of the wild rabbit. It resembles the breast of chicken in color, yet the taste is distinctly and delightfully different. In fact, rabbit meat is in a class by itself when protein tests are made.

Rabbit ranks an easy first in protein, fat and mineral salts over all other meat producing animals.

Compared to other meats, Domestic Rabbit is more easily digested and has greater food value with far less waste. Rabbit meat on the dinner table is as economical, if not more so, than any meat you can purchase. It has very little waste - only 12 to 14 percent of the dressed carcass being waste.

****

NUTRITIONAL VALUES as per U.S.D.A. Circular #549

Edible Portion (Uncooked)	% Protein	% Fat	% Moisture	Calories
RABBIT w/giblets	20.8	10.2	67.9	795
Frying Chicken "	20.0	11.0	67.6	810
Veal (Med.Fat)	19.1	12.0	68.0	840
Turkey w/giblets	20.1	20.2	58.3	1,190
Lamb-Med. fat	15.7	27.7	55.8	1,420
Beef-(Med.Fat)	16.3	26.0	55.0	1,440
Pork (Med.Fat)	11.9	45.0	42.0	2,050

Cholestoral Content-(MG/100 GN Moist Food) compared to other food = Beef (fat) 125, Beef (lean) 95, Veal 125, Chicken (light) 90, Chicken (dark) 60, Pork Spareribs 105, Lamb liver 610, Pork liver 320, Calf liver 360 and RABBIT, a low total of 50.

RABBIT MEAT has the smallest content of uric acid of any meat known.

Inside

WE ARE BREEDERS OF "TOP QUALITY"

White New Zealands

Californians

Champagne d'Argents

AFTER MORE THAN TWENTY-FIVE YEARS OF SHOWING, AND DOING VERY WELL, WE ARE CONTINUING TO BREED IN ACCORDANCE WITH THE A.R.B.A. STANDARDS BUT WITH GREATER EMPHASIS ON

PRODUCTION

LIVEABILITY

GROWTH WEIGHTS

WE LINE BREED ON A SELECTIVE AND COMPLIMENTARY BASIS. MOST OF OUR BREEDERS ARE A.R.B.A. REGISTERED AND ALL FULLY PEDIGREED.

OUR PRICING PROGRAM

WE VERY SELDOM HAVE BREEDING-AGE STOCK FOR SALE.

ON A SELECTIVE AND PRE-ORDER BASIS, WE SELL AT THE RABBITRY AT TWO MONTHS OR MORE OF AGE.

WE DO NOT SHIP

PRICE SCHEDULE

$5.00 per month of age from two to three months. Portions of month are figured.

Over three months of age, pricing subject to negotiation.

We accept orders with cash deposit of $5.00 per head and notify about ten days before date of availability. Deposits are not refundable in cash, only for rabbits or merchandise.

ED-MAR RANCH RAISED/DRESSED RABBITS ARE PRODUCED IN OUR MODERN, CLEAN, SANITARY RABBITRY AND ARE PREPARED AS FOLLOWS:

FRYERS - Whole-bodied, split or cut-up

ROASTERS - Whole bodied

STEWERS - Cut-up

Packaged rabbits are prepared for freezing with net weight and total price on each package. Triple wrapped for your freezer, with a recommended freezer life of three (3) months.

ORDERING

To be sure, we suggest that you place your order on Monday or Tuesday for pick-up on the week-end.

FRYERS are eight to ten weeks of age.

ROASTERS are slightly older but young enough to be tender and flavorful.

STEWERS are usually older rabbits, out of production and weighing from 5-1/2 to 8 lbs.

FROZEN RABBITS

ED-MAR Frozen Rabbits are fresh frozen and are of the same high quality as ED-MAR fresh rabbits. Since, at times, the supply is greater than the immediate demand, it is necessary to freeze some of the product. We will gladly advise regarding several methods for proper defrosting.

Business Card

RABBIT EQUIPMENT SUPPLIES DRESSED RABBITS

SEACORD'S ED-MAR RABBITRY

(A. R. B. A. REG. NO 2275)

REGISTERED WHITE NEW ZEALANDS CALIFORNIANS
CHAMPAGNE D'ARGENTS

EDWARD M. SEACORD
ARBA REGISTRAR

PHONE (408) 378-8976
17015 POLLARD ROAD — LOS GATOS, CALIF. 95030

The by-product business

Rabbit producers have a number of sidelines that can add to profits. These include the pelts of rabbits slaughtered for meat, "lucky rabbit's feet," earthworms and garden fertilizer. All help contribute to the over-all profits of the operation.

Market for pelts

In an earlier chapter we told you how to handle the pelts to preserve them for sale. They are stretched and cured until sufficient numbers are accumulated to attract a buyer. Prices for rabbit pelts vary a great deal along with quality. A summer pelt is likely to be of less value to a furrier than one taken in winter. Recently, prices have been in the 50¢ to 60¢ per pelt range. It's a way to make extra profit from processing.

Here is a list of fur buyers who have purchased rabbit pelts in the past:

American Fur Brokers, Inc.
101 West 31st St.
New York, NY 10001

Boise Basin Furriers
5512 Wylie Lane
Boise, ID 83703

Fillmore Rabbit Co.
669-663 S. Filmore
Allentown, PA 18103

Grand River Fur Exchange
6310 U.S. 6, Dept. R
Rome, OH 44085

Gunter Fur Products
19411 S.E. Green Valley Rd.
Auburn, WA 98002

Johnson Brothers Furriers
342 N. Towne Ave.
Pomona, CA 91767

Bill Prouhet Fur Co.
13032 Gist Rd.
Bridgetown, MO 63044

Quality Fur Dressers (Rex)
5250 San Felipe Rd.
Hollister, CA 95023

4 D Rabbit Farm & Fur Co.
Drawer 288-KA
Huntington, TX 75949

River Rabbit Ranch
c/o Dick Sperlich
Rt. 1, Box 157
Pendleton, OR 97801

Encourage 4-H projects

Rabbits are an official project for 4-H club youth in almost every state. There are thousands of 4-H'ers raising rabbits in major producing states.

Here are the rabbit producers of tomorrow. This is an opportunity for you, as a rabbit producer, to encourage these young people. You not only open up a market for a set of breeder rabbits, you also are introducing another family to the advantages of rabbit meat.

The County Agent, or County Extension Director as he is called in some states, can tell you the names of leaders with members interested in rabbits. Or better yet, ask for an opportunity to meet with the leader and club members. Tell them all about rabbit raising. Take a couple of rabbits along to get them interested. It's good for rabbits in many ways.

Earthworms add income and solve a problem

Many rabbit producers raise earthworms in the dropping pits beneath the hutches. You can get extra income selling these to fishermen if there is fishing in your general area. At the same time, earthworms convert rabbit manure into casts that make it better fertilizer. With plenty of earthworms in the droppings from the hutches, odor problems are greatly reduced. Many owners clean under the hutches only twice a year when earthworms are present.

Build a wooden box about 10 inches high, directly under the rabbits. Make it the same length and width as the cages. Drive stakes into the ground to support the sides. Put in a layer of sawdust to start and the rabbits will do the rest. It may be necessary to sprinkle occasionally to keep the bed moist.

Go to an earthworm supplier to get your beginning stock of worms. The Red Wiggler and the Brown Nosed Angler are popular names for

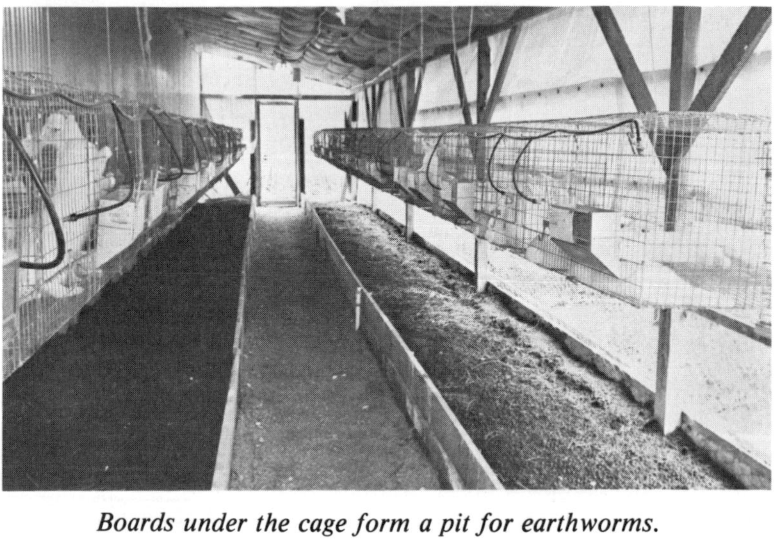

Boards under the cage form a pit for earthworms.

two types of worms that work well in this situation. The worms feed on the droppings which have considerable nutrient value. And they also take care of fines that may fall from the feeder.

The next thing you need is an eager son or someone else who wants to take on the job of selling earthworms to fishermen. A plastic carton of worms can bring a dollar or two and will add to your income.

When you clean out the bed, put the material on the compost pile or spread in the garden. Worms will continue to do good work in both places. Leave enough worms and eggs to reseed the pit under the rabbits when you clean out.

For more information, several booklets on raising earthworms are available. One of them is "Raising Earthworms For Profit" by Earl Shields, Shields Publications, P.O. Box 669, Eagle River, WI 54521.

Sell rabbit manure

It is generally accepted that rabbit manure is among the finest available for gardens. This is particularly true when produced with earthworms. Check around your garden supply store and see how dried cow manure is selling. Rabbit manure is better. You'll have a lot of it, too. A doe and the 40 young she produces each year will produce about eight cubic feet of manure.

If you can't use the manure yourself, sell it by the bushel or ton to local gardners, home owners and truck gardeners.

Rabbit manure has a high nitrogen content when the rabbits are fed a well-balanced diet. It will not burn lawns or plants and is easy to incorporate in the soil. It's satisfactory on gardens and lawns and spread around flowering plants, shrubbery, and trees. There is no danger in using it for fertilizing soil on which crops are to be raised for feeding rabbits.

The value of rabbit manure depends on how it is cared for and used. There will be less loss of fertilizing elements if the material is immediately incorporated into the soil. When manure is stored in piles and exposed to the weather, chemicals are lost through leaching and heat. Much of this loss can be prevented by keeping the manure in a compost heap or in a bin or pit.

Price depends on what you can get, in comparison to other fertilizer, and on local demand. You can expect something in the range of 50¢ to $1.00 per bushel. One grower is satisfied with $15 per pick-up load.

Novelties add income

You can get top dollar from rabbit pelts by converting them into novelties. These include clothing, decorative accents, toys, and dozens of other items. Look around at flea markets, fairs and art shows. Dream up designs you think will sell and give them a try. If your wife has a flair for this kind of thing, novelties can be a very profitable part of the rabbit operation. Some items often made are: key charms, tassels for clothing, muffs, mittens, stuffed toys, and many other items.

"Lucky" rabbit's feet are always in demand. Here's how to prepare them:

Mix a solution of 20 percent alcohol, 20 percent formaldehyde and 60 percent water. Place the feet in this solution for approximately five days and then let them dry in a well-ventilated area. Then dye the feet if you choose, and put the metal cap and chain on.

Build volume—sell anything you can

Stock up on supplies and equipment. When you sell rabbits, try to sell everything else the buyer needs. Keep a few hutches, waterers, feeders, etc. on hand. Mark them up to provide a good profit. We've already recommended that you have bags of feed ready to go. Stress the need to keep the rabbits on the same ration and you'll make the sale. The new customer is concerned about care for his rabbits. He wants all the advice he can get. He needs to buy his equipment somewhere...why not from you?

Good luck in your rabbit business

There's no easy, sure-fire way to develop a profitable rabbit business. The first key is efficient production of big litters on a regular schedule. Feed them right to finish on 4 pounds of feed per pound of gain, or less.

The next big challenge is marketing. No single market is likely to be as profitable as a combination. Sell in several different ways. Don't depend on a single market or product. To date, rabbit meat outlets have been up and down. The processor you work with today may change the rules or be gone tomorrow.

There are only so many dollars in a rabbit. The more of them you can keep for yourself, through your own marketing efforts, the better off you will be.

Growers love to work with rabbits. They enjoy raising them and in talking rabbits with other breeders. Shows and other events are fun. Many knowledgeable people say rabbit production today is where broilers were 40 years ago. We all hope that more and more Americans will learn the value of tasty, nutritious rabbit. That's what it takes to build the demand that cries out for stocking at the supermarket.

These New Zealand White fryers are ready for market

Chapter 8

Harvesting your rabbit crop

Rabbits are born, grown out and ready for harvest in just eight weeks. Look upon your rabbits as a crop, like that which grows in your garden. Your plants flourish for a brief season, then are harvested to feed the family or others.

Slaughtering a rabbit is very much like killing a chicken or dressing a duck or deer shot in the wild. No one will find the task pleasant but after the first rabbit or two, processing at home will become routine for you.

Unless you find a market that buys live rabbits, you will have to learn to do the processing for yourself. It doesn't make economic sense to hire someone to do the job if your goal is meat for the family. And if you sell the crop from the starter unit, chances are you will find it necessary to do the butchering for your customers.

Your goal must be a quick painless kill, followed by an efficient dressing routine that preserves the quality of the meat. Pick out a spot in the basement or in an outbuilding where water is available and get organized.

You'll need hooks to hang the rabbits while you are dressing them. Have the necessary knives, sharpened and ready to go to work. Be prepared to handle the meat in a sanitary manner.

Build the rabbitry slaughter in a clean location, where it can be sanitary and free of flies and dusts. For most small set-ups, the basement of the home is the logical location.

Equipment you will need

The tools for dressing rabbits are simple. Nail a board to a wall at about head level. Screw, in a horizontal row, six hooks with sharpened points, about 8 inches apart. A small table about 30 inches high should stand near the board.

On the table have two large buckets of cool, clear water. One bucket is for washing blood from your hands, knives, and equipment, and the other is for washing and cooling the dressed rabbits.

Under the hooks you need a washtub to catch the blood, intestines, heads and feet. Another good system is to hang a large heavy-duty plastic bag to the hooks. Entrails then fall directly into the bag.

Two knives are needed, a chicken-sticking knife (slender blade) and a boning knife (heavy blade). Keep them sharp.

Have several wire hide stretchers ready. On the table have a coffee can of clothes pins for fastening the hides to the stretchers.

Have a pan or some other clean container ready for the meat.

Get some help at first

Although we have some excellent "how to" pictures on the following pages, you probably can use some help from an experienced person at first. Learn how to cut-up a rabbit properly at the beginning. People are particular about the food they buy and the way you cut the rabbit can have a lot to do with sales...and repeat business.

The steps we show here have been established by those with long practice. After a few trials, you'll be able to do the job routinely in five minutes or less.

The steps in rabbit dressing

There are two generally accepted ways to kill a rabbit. One is to stun the rabbit by striking with a stick at the base of the skull behind the ears. Immediately suspend the rabbit and remove the head for bleeding.

The other method is to dislocate the neck. Again the rabbit is hung and bled immediately. It takes some skill to break the rabbit's neck in the manner shown in the pictures at right. It is a matter of learning how and then trying it for yourself a few times.

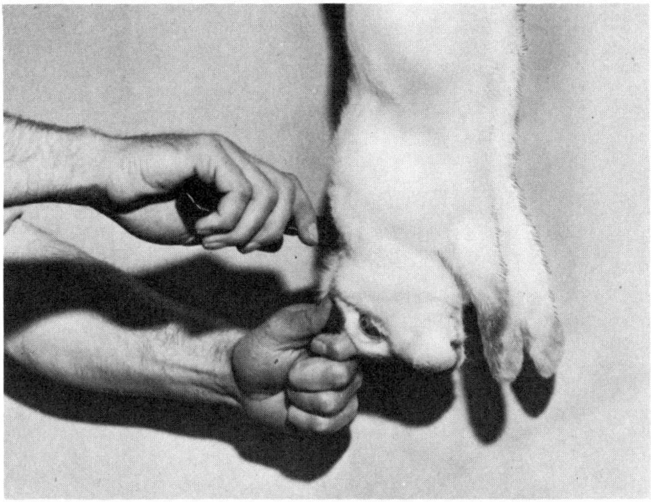

Step 1 To dislocate the neck, hold the animal by its hind legs with the left hand. Place the thumb of the right hand on the neck just back of the ears and place the fingers under the chin. Stretch the animal by pushing down on the neck with the right hand. Press down with the thumb, then raise the animal's head with a quick movement to dislocate the neck. This method is instantaneous and painless.

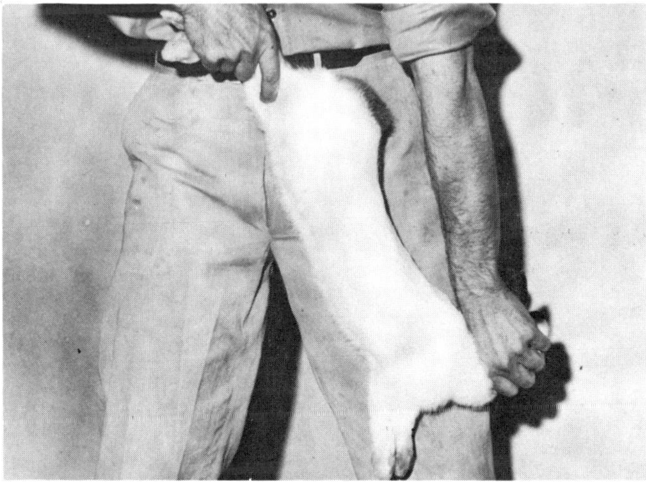

Step 2 Hang the rabbit on the wall hooks by the left hind leg. Insert the wall hook between the tendon and bone of the leg. Immediately cut off the head. Cut close to the head through the place where the head was broken away from the neck.

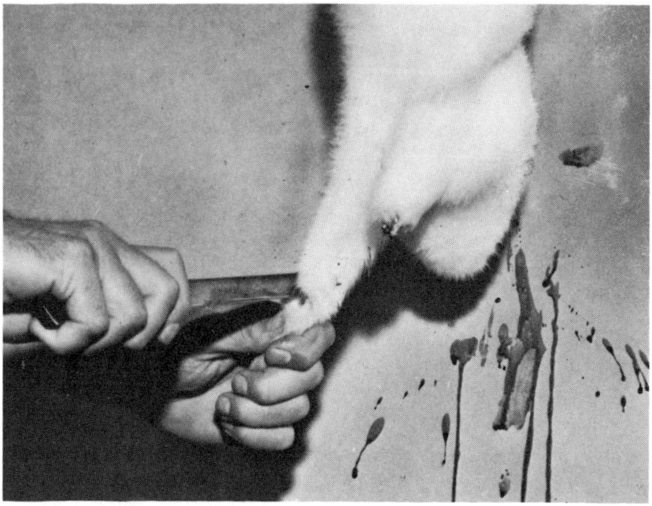

Step 3 Use the boning knife to cut off both front feet, cutting through at the knee joint.

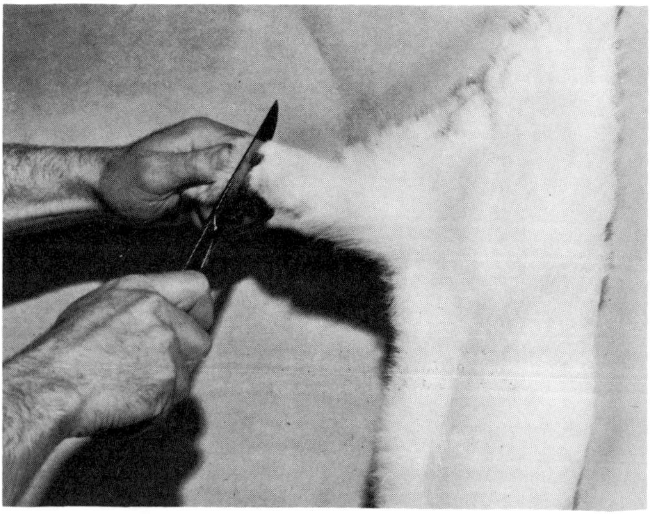

Step 4 Cut off the hind foot that has not been used in hanging the rabbit.

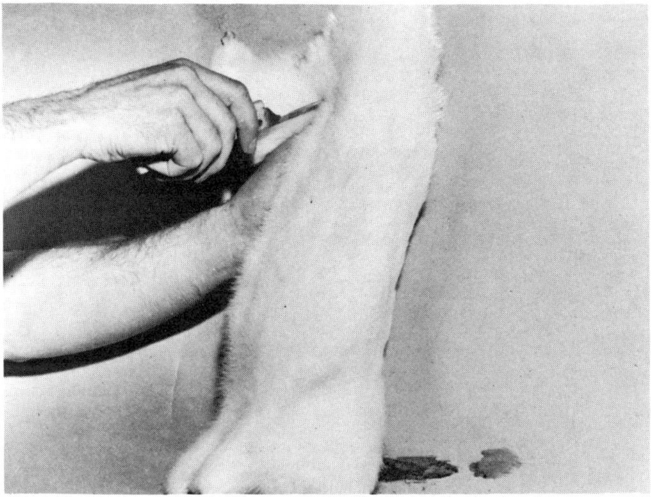

Step 5 Use the chicken sticking knife to slit the skin up the inside of both hind legs.

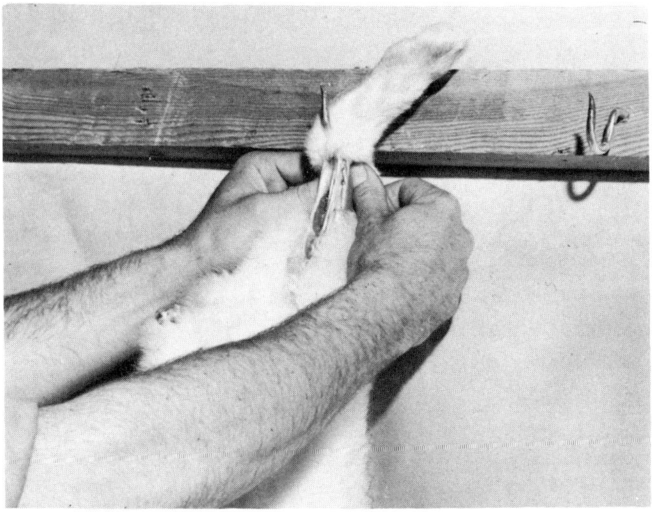

Step 6 Tear away the hide from the hind leg that is suspended on the hook.

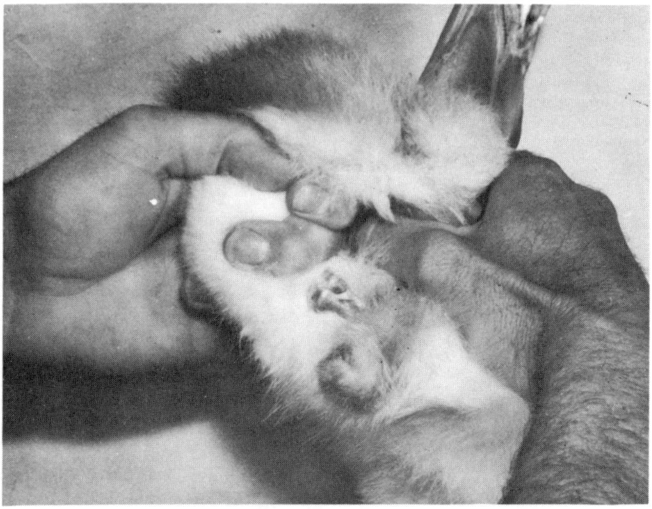

Step 7 Tear the hide from the tail and vent by working fingers between the hide and body ahead of the tail and over the rump.

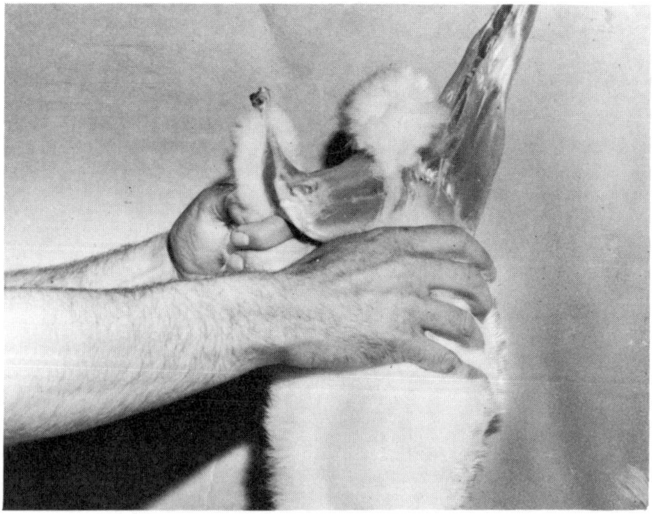

Step 8 Force your fingers between the hide and body and pull the hide from the free hind leg.

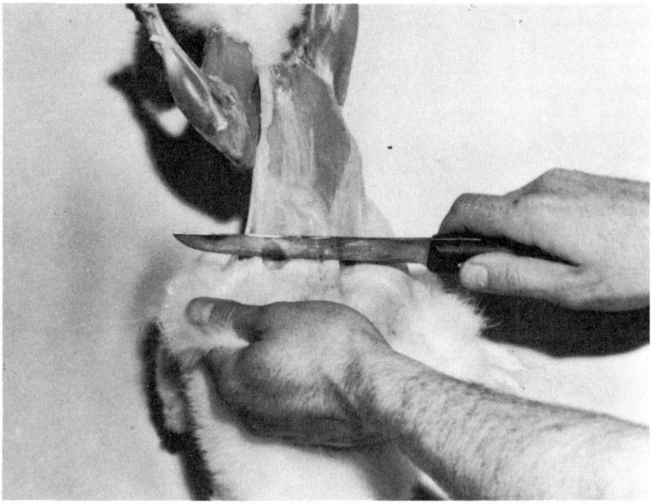

Step 9 Cut as shown here. Leave the fat on the flanks, not on the pelt.

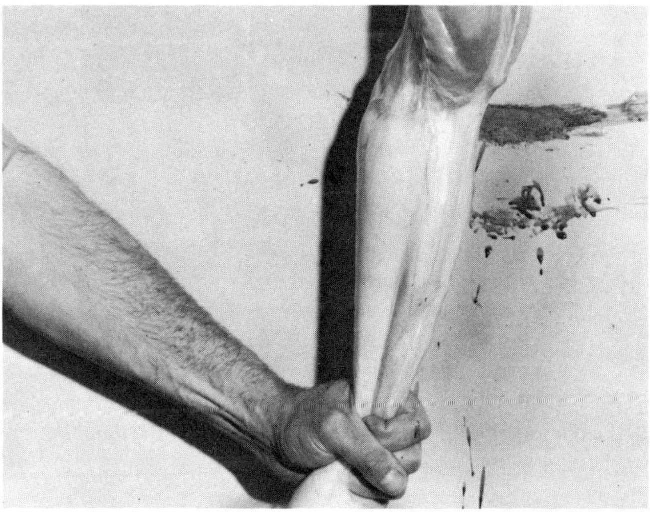

Step 10 As soon as you can hold the entire pelt with one hand, remove it with one strong pull, like you peel off a sweater. Dry the skins indoors.

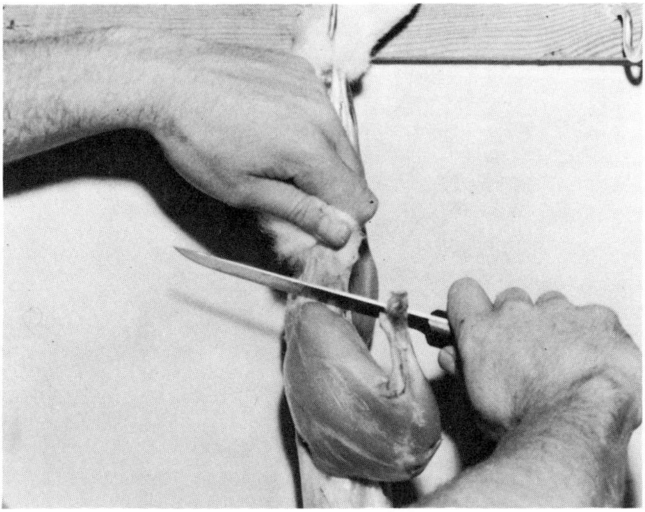

Step 11 Cut off the tail where it is attached to the rump. Save it if you have a market for bunny tails.

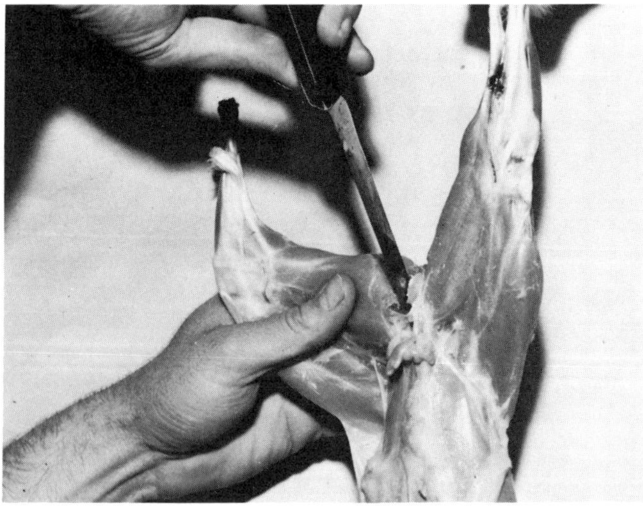

Step 12 Cut the pelvic bone between the hind legs by inserting a knife from above and prying out.

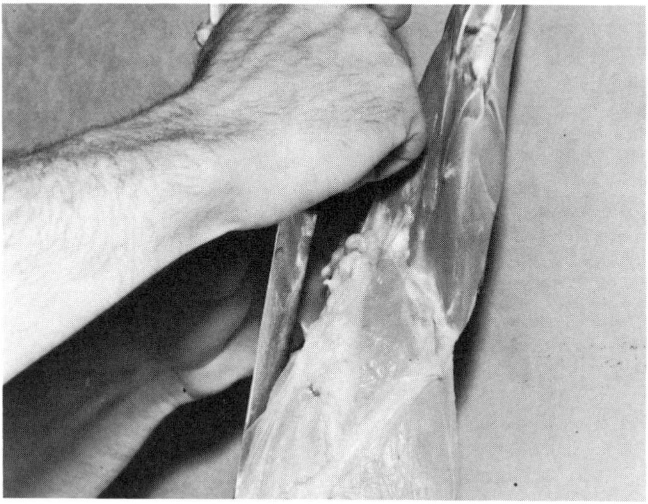

Step 13 Slit down the belly. Be very careful not to cut into the bladder, intestines or stomach. This could contaminate the meat.

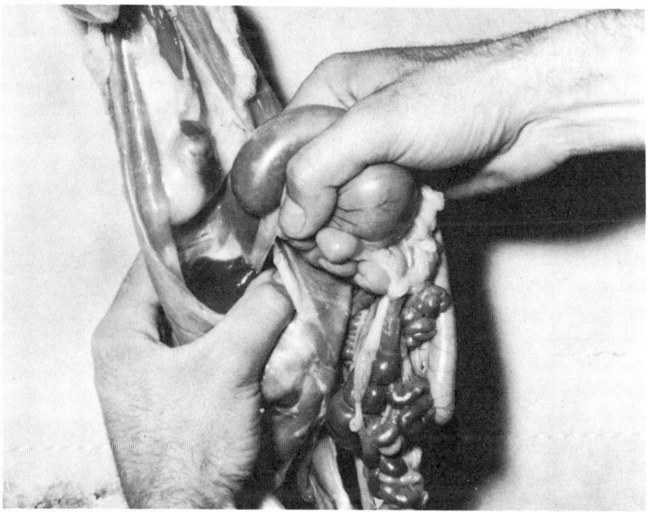

Step 14 Pull out insides by grasping the stomach and holding liver in place with thumb of other hand. Leave kidneys, liver, heart and lungs in place.

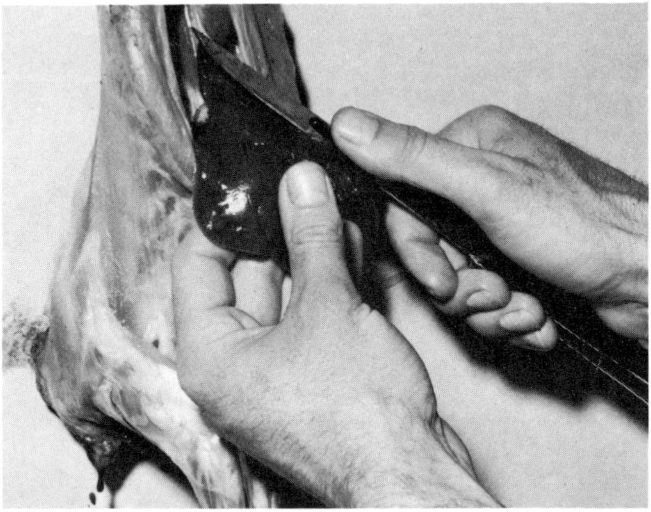

Step 15 Carefully remove the gall bladder without cutting or breaking it. The bitter green bile of the gall bladder must not be spilled on the meat.

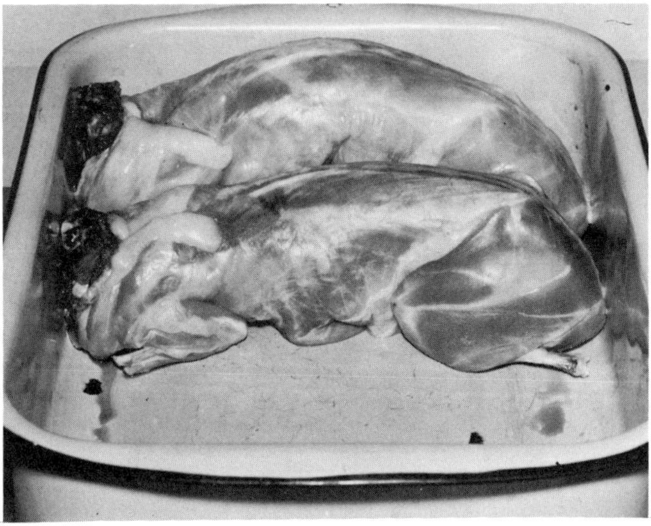

Step 16 Wash the carcass in cold water. Leave it in the cold water for about 15 minutes. Remove and place in a pan or wire basket in a natural position.

Use care in slaughter so hairs will not get on the carcass. They are difficult to remove and give it an unattractive appearance. Rinsing in cold water, as recommended, helps remove hair and blood, and also cleans the carcass.

Do not leave the carcass in the water for more than 15 minutes. Prolonged soaking allows the meat to absorb water. This is considered an adulturation of the product. After removing from the water, chill the carcass in the refrigerator. Try to keep it between 36° and 40° F for 24 hours. Then cut up for cooking at home or for sale.

Rabbit cut into eight pieces

This picture shows the customary way to cut up a rabbit. You wind up with seven or eight good pieces of rabbit.

Steps in cutting up a rabbit

Cutting up a rabbit is much like the system used in preparing a chicken for the frying pan. Basically, the meat is cut at joints to make seven or eight nice pieces, plus liver and heart.

1. Start with a front leg. Pull it out, then use a sharp knife, cut to the shoulder joint. Twisting the leg helps you locate the joint. Repeat on the other front leg.

2. Pull out a rear leg and cut where the thigh joins at the hip joint. Again, twist to help locate the joint. The leg is not separated from the thigh when cutting up a rabbit. Repeat on the other leg.

3. Cut away the meat where the rib cage ends. Then bend the spine backwards until it snaps. Cut through with the knife.

4. Slice along the cartilage that joins the ribs and backbone along the spine. This makes two pieces of rib.

5. Cut the rear section of the back if desired. This is the loin of saddle and is considered the real delicacy of the rabbit.

Two ways to package a rabbit for the market

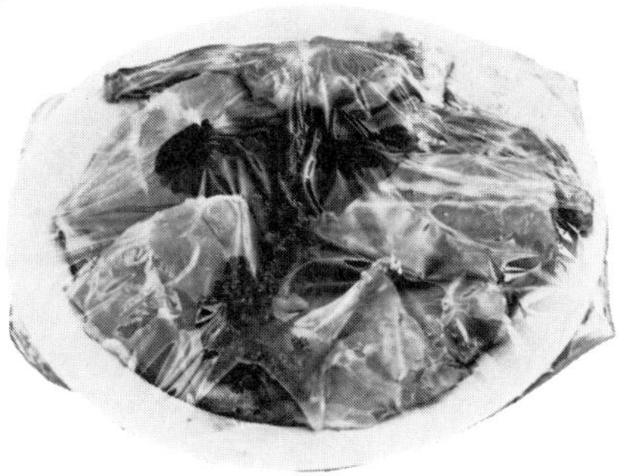

Packaging for sale

Hotels, restaurants and other commercial buyers probably will want to buy the whole carcass. But for sale to local customers, you'll want to do a good job of packaging.

Housewives usually prefer the cut-up packaged product. Use a knife to cut up the carcass. Do not use a cleaver because it may splinter the bones.

A paraffined box with a cellophane window makes a neat, sanitary package for the chilled rabbit carcass. If the meat is to be frozen, wrap the box in a special sealable wrapping to prevent freezer burns and loss in palatability.

A box 9 inches long, 4 inches wide, and 2½ inches deep is suitable for a fryer carcass weighing 1¾ to 2½ pounds. Arrange the cuts attractively. Include the heart, kidneys, and liver.

If you sell to the home trade or furnish butchers with meat that is to be consumed locally, you can make a neat, sanitary, and inexpensive package by arranging the pieces and a sprig of parsley on a paper plate. Cover the plate with a piece of clear cellophane.

Know your state's meat laws

If you are going to sell processed rabbit meat, it will be necessary to comply with your state's meat inspection laws. Most states require that rabbits meet specified minimum standards. If you do not know who does the inspection in your state, check with a local butcher or meat market. He can help you determine the requirements.

Many small producers sell the live rabbit to a customer, then butcher it for him. This sidesteps the processing regulations in most states. But if a customer is going to watch you do the work, you'll still need high standards of sanitation.

The pelts have some value

Most rabbit producers find it pays to preserve the pelts. The best way to make them profitable is to produce a garment or some novelty item yourself. Pay from commercial sources is low...but it does bring in income. It takes large volume to attract a buyer.

Over 100 million rabbit skins are used by garment manufacturers in this country every year. However, most of them are imported. We are concentrating on the 8-week rabbit weighing around four pounds. This produces what the fur trade considers a small skin. The skin is not as mature as the trade would like and is not as durable.

How to cure and tan pelts

Most pelts have some value in the fur trade. If you butcher your own rabbits, you can add to your income by properly caring for your pelts and marketing them. Or, you may wish to tan your own pelts.

You can cure the pelts or tan them. To cure the pelts, while still warm, place skins flesh-side out, on wire or board, formers or shapers (with the fore-part over the narrow end).

You can make a satisfactory skin shaper from 5 feet of number 9 galvanized wire. It is not desirable to stretch the skin unduly, but to hold it in place. Mount a skin on the shape, making sure both front feet casings are on the same side. Take care to remove all wrinkles. Scrape all the fat from the hide, being careful not to tear the hide. Be sure skins are thoroughly dry, but do not dry them in the sun or with artificial heat. If they are going to be stored, sprinkle with naptha flakes. Never use salt in curing rabbit skins.

You can tan your own pelts by the following process.

1. Remove pelt from rabbit and split down the belly.
2. Cut off front legs and open to outer edge of skin.

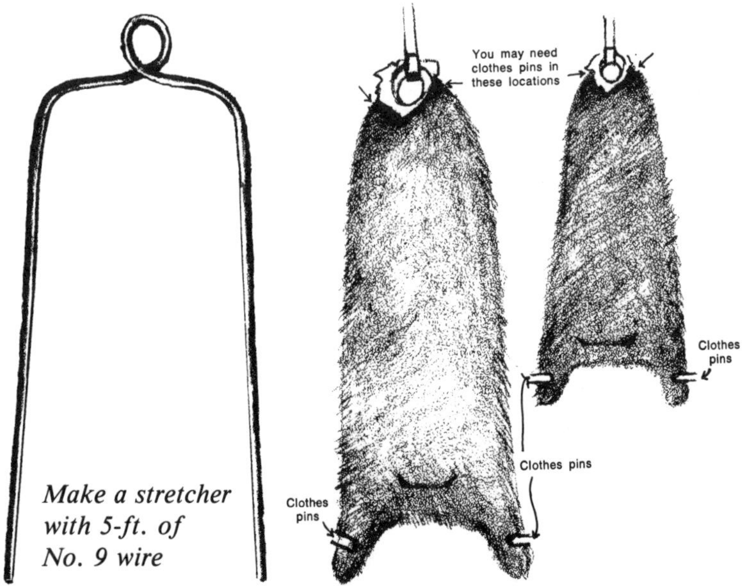

3. Remove any excess fat. If the fresh pelt cannot be used that day, place in a plastic bag and freeze until needed.

4. If dried already, soak in water until soft and use like the fresh hide.

5. Place skins in one of the following solutions made in a plastic pail with a lid:

2 gallon of water		2 gallons of water
2 ounces sulfuric acid	OR	8 ounces battery acid
2½ pounds salt		2½ pounds salt

In making the solution, place the salt in the water. Mix well. Then slowly pour the acid into the solution. (Do *not* pour water into acid.) Acid will burn. Once the solution is made, with dilution suggested above, it is safe to hands.

6. Leave 24 hours or more.

7. Remove and pull flesh from the skin, working from the tail end.

8. Return to the solution after 48 hours or more. Skin will no longer feel slimy.

9. Remove and wash in a mild detergent. Rinse well and squeeze out excess water.

10. Let dry slowly. As they dry, stretch to break the fibers. This will cause the skin to turn white and will soften it. The more it is pulled and rubbed, the softer it will become.

11. Try to stretch the pelt flat as the drying is completed.

Chapter 9

Tasty meals featuring rabbit

Here is a taste sampler of mouth-watering rabbit dishes you'll want to try. Some are continental recipes handed down through the generations. Others are modern-day versions demanding a minimum of time and effort.

We believe the key to getting people to eat more rabbit in this country is to encourage them to try it! Once tasted, Americans will learn what Europeans have known for centuries... rabbit is truly a delicacy. And what a boon for health-conscious dieters! Higher in protein than any other meat, yet lower in calories, it also boasts less fat, uric acid, cholesterol and sodium than any other meat available today.

As with any quality meat, flavor begins with proper care and feeding of the animal. The feeding system developed for today's domestic rabbit meets this criteria. Serving a tender young fryer, prepared with skill and respect, will do wonders in popularizing rabbits as a food throughout America.

In this chapter, we share with you recipes from many sources including several graciously contributed by rabbitry owners. We wish you, your family and friends, many delicious forkfuls!

Rabbit with Rice

A favorite gourmet treat

1 fryer rabbit, cut up
½ lb. butter or margarine
2 cups fresh sliced mushrooms
2 cups cream
1 cup beef or chicken bouillon
½ cup white wine
1 cup rice
1 cup frozen peas
3 Tb. pimiento
Salt and pepper to taste

Saute rabbit in the butter until golden brown; remove from pan. Saute mushrooms in remaining butter; add cream, bouillon and wine to the mushrooms, stirring constantly. Set aside. Bring to boil 1¾ cups water, add ¾ tsp. salt; stir in one cup rice. Cover and cook over slow heat approximately 30 minutes. Uncover the pot for the last 5 minutes and continue to cook over slow heat, shaking the pot from time to time until grains are separated. Cook 1 cup of frozen peas and add to rice; add 3 Tb. chopped pimiento.

To serve, place a mound of rice on plate, top with serving of rabbit, spoon mushroom sauce over all. Garnish with chopped parsley.

Sawley G. Wilde, Wilde Hare Rabbitry, Sheridan, Wyo.

Board Valley Rabbit Bar-B-Q

Serve hot with hamburger rolls

2 or 3 large rabbits
1 cup of catsup
5 Tb. of Worcestershire sauce
1 small bottle barbeque sauce
4 Tb. chili powder
3 large onions, diced
3 large green peppers, diced
5 to 6 stalks of celery, diced
1 clove of garlic, diced
Salt and pepper to taste

Boil rabbit until tender enough to remove meat from bone. After removing meat from bone, dice into small pieces. Using large pot, add all ingredients and slow-cook for about two hours, stirring occasionally.

Mrs. Diane K. Ronchi, Board Valley Rabbitry, Sparta, Tenn.

Board Valley Baked Rabbit

A Tennessee favorite!

1 or 2 fryer-sized rabbits
Salt and pepper
Ground thyme
Bay leaves
1 lb. bacon
Seasoned bread crumbs
1 cup water
¼ lb. melted butter
2 diced onions

Cut rabbits into serving pieces. Soak 12 to 24 hours in salt water, in refrigerator. After soaking, pat pieces dry and place in buttered baking dish. Sprinkle with salt, pepper and ground thyme. Add several large bay leaves and diced onions. Top off with either diced or whole slices of bacon. Pour remaining water in which you soaked the rabbit into bottom of pan, along with butter. Cover with foil, and bake at 350 degrees until tender. Remove cover and sprinkle 1 cup of seasoned bread crumbs on top of rabbit mixture. Bake about 20 minutes more.

Board Valley Baked Rabbit With Mushrooms

"Super Simple!" "Tastes Fantastic!"

1 fryer-sized rabbit, cut into serving pieces
2 cans of cream of mushroom soup
Salt and pepper to taste
½ stick of butter, melted
½ lb. fresh mushrooms, sliced

Salt and pepper rabbit pieces to taste. Place pieces into a shallow baking dish, pour melted butter over all and bake, covered, in a 350 degree oven for 30 minutes. After 30 minutes, remove cover. Add soup and fresh mushrooms and bake another hour, or until tender.

Mrs. Diane K. Ronchi, Board Valley Rabbitry, Sparta, Tenn.

Pot-Roasted Rabbit

Fill your kitchen with a spicy fragrance all day long

1 4-6 lb. rabbit, cut up
Salt and pepper to taste
Flour for dredging
1 clove garlic, sliced

2 medium onions, sliced
2 bay leaves
1 whole clove
1 cup hot water
2 Tb. soy sauce

Rub all rabbit pieces with salt and pepper. Dredge lightly in flour. In skillet, cook the rabbit pieces to a good golden brown. Place 1 sliced onion in the bottom of the crock pot. Add rabbit and remaining ingredients and cover. Cook 10 to 12 hours on low heat. Thicken gravy after rabbit is removed.

Rabbit Creole

A delicious Creole flavor for rabbit prepared in a slow cooker

1 fryer rabbit, cut up
½ cup flour
2 t. salt
½ cup butter or margarine
½ cup chopped onion
¼ cup minced green pepper
1 - 1# can tomatoes
¼ cup chopped celery
½ cup catsup
¾ cup water
2 t. Worcestershire sauce
½ t. chili powder
Dash pepper
2 Tb. butter or margarine
2 cups cooked rice
1 - 4 oz. can mushrooms

Dredge rabbit in flour mixed with salt. Brown in butter or margarine in skillet. Remove rabbit, saute onion, celery and green pepper in remaining butter or margarine. Add tomatoes, catsup, water, Worcestershire sauce, chili powder and pepper to vegetables. Stir to blend. Place rabbit in the slow cooker, pour sauce over rabbit. Cover and cook on low heat 6 to 8 hours. Saute mushrooms in butter in fry pan, and add to rabbit before serving. Serve sauce over cooked rice.

Seacord's Ed-Mar Rabbitry, Los Gatos, Calif.

Roast Rabbit With Stuffing

A tantalizing combination of flavors

1 whole rabbit

Mix ingredients for dressing (recipe follows), and stuff in rabbit. Sew stomach cavity closed. Place in 400 degree oven covered tightly with foil, for 1 hour, basting every 30 minutes. Use additional butter if needed.

Remove foil and bake another 30 minutes to 1 hour, depending on size of rabbit. Baste every 15 minutes.

Variation: Use barbeque sauce for basting.

For Dressing:

- Fresh bread, torn into small pieces
- 1 can mushrooms and juice
- 1 or 2 stalks celery, sliced
- ½ t. poultry seasoning
- ½ t. salt
- Dash pepper
- 3 Tb. butter

Roger & Roberta Draper, High Country Rabbitry, Magna, Utah

Hawaiian Rabbit

A sweet-sour inspiration reminiscent of the Islands

- 1 frying rabbit
- Flour, salt, pepper
- 1 cup unsweetened pineapple juice
- ¼ cup white vinegar
- 1 cup pineapple chunks
- 1 green pepper, cut up
- 2 Tb. fat
- 1½ Tb. cornstarch
- ¼ cup sugar
- ½ cup water

Wash rabbit well, and cut up. Mix flour, salt and pepper. Roll rabbit in mixture. Melt fat in skillet, brown meat on all sides. Add pineapple juice and vinegar. Simmer 40 minutes. Add pepper and cubed pineapple; cook 5 minutes. Remove meat to hot platter. Mix cornstarch and sugar together, add water, and add to skillet, cooking sauce until thick. Serve over rabbit. (4 servings)

Lydia H. Amann, Burlington, Wisc.

Minnesota Baked Rabbit

Easy and quick to prepare... but oh, so good!

1 rabbit browned in a mixture of 1 t. paprika, ½ t. tumeric, ½ t. garlic salt, ½ t. onion salt and flour.

Put into casserole. Add 1 med. sliced onion. Bake at 375 degrees for 45 minutes, (covered).

Combine 1 can celery soup and ⅔ cup light cream, pour over meat. Bake 20 minutes more, uncovered.

Serves 4.

Mrs. Robert Bowe, Waterville, Minn.

Oven Baked Rabbit with Sour Cream Gravy

Tender and delectable

2-(2 lb.) rabbits or 1 (4-5 lb.) rabbit, cut up
Flour for dredging
¼ cup oil
2 Tb. butter
1 cup onion, chopped
1 clove garlic, minced
2 t. salt
¼ t. pepper
1 cup milk
1 cup sour cream

Coat rabbit with flour (about 6 Tb.). Heat oil and butter in ovenproof skillet, add rabbit and cook until golden brown. Remove to hot platter. Cook onion and garlic in same drippings until tender. Stir in 1 Tb. flour, salt and pepper. Gradually blend in milk, stirring constantly, until mixture is thickened.

Return rabbit pieces to skillet, spoon sauce over pieces. Cover and bake 1 to 1½ hours, or until meat is tender. Remove pieces to hot platter. Stir sour cream into gravy, stirring constantly until hot. Serve over rabbit. (Serves 8)

Rabbit Italian Style

An exciting flavor from the continent

1 medium-sized rabbit
½ cup oil
½ cup red wine vinegar
1 medium onion, sliced
1 t. thyme
1 clove of garlic, crushed
¾ cup butter or other fat
2 carrots, sliced
1 cup whole button mushrooms
3 medium-sized tomatoes, chopped (or 2 c. canned)
1 bunch green onions, chopped
2 cups red table wine
Salt and pepper to taste

Cut rabbit into pieces: Pour over it the oil, vinegar, onion, garlic, thyme, salt and pepper and allow to marinate for 24 hours. The following day, melt butter or fat. Add the tomatoes, onions, mushrooms, and carrots and simmer for 5 minutes. Then add rabbit, and its marinade. Pour red table wine over all, cover tightly, and cook slowly until tender (about one hour).

Rabbit Stew

Vary the seasonings for fun and variety!

1 rabbit, cut in pieces
Marinate for 12 hours in white wine.

	Seasonings
4 slices bacon	2 bay leaves
1 large onion, minced	1 sprig parsley
1 clove garlic, crushed	1 t. savory
12 mushrooms, sliced	1 t. marjoram
Juice of half a lemon	1 t. salt
1½ cup marinade, strained	3 peppercorns
1 cup water	

Put bacon and onions in earthern pot. Arrange on top, pieces of meat. Add garlic and seasonings. Sprinkle juice of lemon over mushrooms and add. Mix 1½ cups of strained marinade with 1 cup water. Pour liquids over all, cover pot, and bake 3 hours at 300 degrees.

Old World Hasenpfeffer

A European specialty for centuries

1 large rabbit (or 2 small rabbits) cut up
1 cup white distilled vinegar
1 cup beer (or water)
1 sliced onion
1 small stalk celery
1 ounce (2 Tb.) whole mixed pickling spices
½ t. coarsely ground pepper
1 t. salt
½ t. thyme leaves
2 bay leaves
3 slices bacon
1 Tb. sugar
3 Tb. flour
2 cups marinade
3 old-fashioned gingersnaps
½ cup sour cream

Place rabbit in large earthenware, glass or enamel bowl. Combine vinegar, beer, sliced onion, celery, spices, salt and pepper, thyme and bay leaves. Pour over meat. Place in refrigerator 12 to 24 hours, turning occasionally. Remove rabbit; pat dry. Dredge in flour.

Cook bacon in heavy skillet, remove from fat, and set aside. Add rabbit pieces and brown well (add more butter if needed). Sprinkle with sugar, cover and cook until tender, about 1 hour. Strain marinade, and add 3 to 4 Tbs. to skillet if needed. Remove rabbit; keep warm.

Add 3 Tbs. flour to drippings in pan, and add 2 cups of strained marinade together with crumbled gingersnaps. Cook until thickened. Add sour cream, rabbit and bacon pieces. Serve hot.

Shake and Bake Rabbit

The secret of serving tender rabbit: steam for at least one-half the cooking time.

1 cut-up rabbit
1 package of Shake and Bake for Chicken
 or Barbeque flavor Shake and Bake

Prepare rabbit the same as directed for preparing chicken. Bake in a foil-lined pan, covered with foil, for one hour. After one hour of steaming, remove top foil and crisp for 15 minutes.

Barbequed Rabbit

1 cup-up rabbit
1 bottle smoke-flavored barbeque sauce
Seasoning salt
Lemon pepper

Line shallow baking pan with foil. Arrange rabbit in single layer in pan. Sprinkle with seasoning salt and lemon pepper, to taste. Pour a little smoke-flavored barbeque sauce on each piece of rabbit.

Cover tightly with foil, and bake in 400 degree oven for 45 minutes. Remove foil, baste and turn rabbit pieces and bake an additional 30 to 45 minutes, basting every 15 minutes.

Roger and Roberta Draper, High Country Rabbitry, Magna, Utah.

Country Casserole

Gourmet results with minimum effort

1 small rabbit, cut up
3 Tb. cooking oil
1 onion, chopped
¼ lb. bacon, diced
Flour for dredging
Salt

Black pepper (freshly ground)
1 Tb. tomato paste
2 cups dry hard cider
1 bouquet garni *(see "Rabbit Accents" page 143)*
3 medium sized carrots, sliced

Heat oil in skillet or flameproof casserole. Add onion and bacon. Fry until golden. Coat rabbit in flour seasoned with salt and pepper. Add to casserole, and brown on all sides. Stir in tomato paste and cider. Bring slowly to boil, stirring constantly. Add bouquet garni of your choice, and carrots. Cover and bake in 350 degree oven for 1 to 2 hours, or until rabbit is tender. Discard bouquet garni before serving. Season to taste.

Rabbit with Prunes

A French provincial standard.

1 rabbit
Flour for dusting
1 onion, chopped
2 to 4 Tb. butter
1 cup red wine
½ lb. dried prunes, soaked overnight in just enough water to cover
Salt
Freshly ground pepper

Marinade:
1¾ cup red wine
1 cup wine vinegar
Salt
6 black peppercorns
½ t. thyme
1 bay leaf

Mix ingredients for marinade in bowl, add rabbit, and soak for 12 hours, turning once or twice. Pat rabbit dry, and cut into joints. Sprinkle pieces well with flour. Melt butter in saute pan, brown rabbit. Remove meat to platter, and saute onion in same butter. Return pieces of rabbit to pan and add red wine, the prunes in their liquid, salt and pepper. Cover pan and simmer for 30 minutes, turning pieces occasionally. Uncover and simmer 30 minutes more or until meat is tender. Season to taste. Serves 4.

Busy-Day Rabbit with Vegetables

Use your Pressure Cooker to prepare a delicious dinner in a hurry!

2 rabbits, cut up (3 to 4 lb. each)
¼ cup flour
¼ t. pepper, freshly ground
2 t. salt
1-2 t. curry powder if desired
⅛ t. paprika
Bacon fat for frying
6-8 potatoes, peeled
8-10 whole carrots, peeled
1-2 onions, sliced
1 cup hot water
1 bay leaf
1 cup dairy sour cream

Dredge rabbit pieces in mixture of flour, salt and pepper. Brown meat until golden brown in hot fat, in pressure pan. Add vegetables, water, and bay leaf. Cover and cook at 10 lbs. pressure 18 minutes. Reduce pressure slowly. Arrange rabbits and whole vegetables on heated platter. Put remaining vegetables in pan through a food mill or sieve, return this mixture to pan, and stir in cream. Simmer 5 minutes. Serve over rabbit and whole vegetables. (8 servings)

Company Fricassee

A subtle combination of flavorings and spices

1 rabbit

¼ t. each of: cloves, nutmeg, mace, allspice, salt and black pepper. Rub into rabbit meat, let stand for one hour, then cut into serving pieces.

½ cup butter
½ cup chicken stock
½ cup white wine
Yolks of 3 eggs
½ cup cream

Brown rabbit pieces in melted butter, in heavy Dutch oven or pan. Add stock and white wine, and let simmer until tender. Remove meat to heated platter. Meanwhile, beat egg yolks well. Add cream to gravy, bring to a simmer, and stir in eggs. Do not boil. Pour over meat. Sprinkle with parsley to serve.

Fried Rabbit

Serve with Brown Gravy or a flavorful cream sauce

1 fryer rabbit, cut up
¼ t. pepper
1 slightly beaten egg
1 t. salt
1 cup dried bread crumbs
2 Tb. water

Wash and wipe rabbit with damp cloth. Cut into pieces for serving. Beat egg slightly, and season with salt and pepper. Dip rabbit pieces into egg mixture, then roll in crumbs. Brown in ¼" hot butter. Reduce heat. Cover and cook slowly until tender (30 minutes to 1 hour). Serve with gravy.

Brown Gravy

Measure 4 Tbs. of butter left in skillet, draining off remainder. Blend in 4 Tbs. flour with salt and pepper to taste. Cook until brown, stirring constantly. Stir in 2 cups water. Continue to stir and cook until thickened, about one minute.

Deluxe Cream Sauce for Rabbit

2 Tb. flour
2 Tb. butter
1 cup heavy cream (use chicken broth for
part of liquid if preferred)
2 Tb. sherry
Juice of 1 lemon

Mix together flour and butter to form paste. Slowly add the cream, stirring until thickened. Blend in sherry and lemon juice. Add salt. Pour over hot rabbit pieces on platter.

Rabbit Soup

A great way to capitalize on leftover rabbit stock

To liquid remaining from cooking rabbit, add vegetables of your choice such as potatoes, carrots, celery, cabbage. Simmer and add canned tomatoes, salt, pepper, seasonings of your choice, and simmer until done (approximately 3 hours).

For Quickie Rabbit Soup, use stock left from cooking rabbit and add enough additional liquid to make 7 cups. Add Chicken Soup Starter, and cook according to directions.

Janie Bennett, Marana, Ariz.

Rabbit Salad Sandwiches

Add zip to your next luncheon!

1 4-lb. rabbit, dressed
4 eggs
Mayonnaise
Sweet pickles
Salt

Cook rabbit in covered pot with about 2 cups water until done (1-4 hours). Add eggs for the last 10 minutes, to hard boil. Bone rabbit and cut or tear into small pieces. Dice pickles and cooled eggs. Combine and add mayonnaise, using enough to make creamy. Add salt to taste. Serve on toasted bread. Save the cooking liquid to make soup.

Two B's Bar B Q

The secret's in the sauce

1 fryer or roaster rabbit
½ t. garlic salt
½ t. onion salt
½ t. paprika
¼ t. black pepper
½ cup flour
1 cup lard or cooking oil

Cut rabbit in desired pieces (we choose 4 legs, 3 backs, and 2 without bone off rib cage). Sprinkle or shake garlic salt, onion salt, paprika and pepper on both sides of each piece. Dip separately in flour, and brown slowly in oil. Place in shallow pan for oven baking. Mix barbeque ingredients, pour over browned rabbit and bake at 350 degrees 35 minutes, or until tender.

Bar B Q Sauce

1 cup diced onion
¼ cup diced bell or sweet pepper
¼ cup celery
1 Tb. Worcestershire sauce
½ t. prepared horseradish
1 t. vinegar
1 t. prepared mustard
1 Tb. sorghum, brown sugar, or dark syrup
2 6-oz. cans tomato paste

Bunny Lady and Bill Cox, Locust Grove, Okla.

Dad's Rabbit

Holds well in oven for latecomers

1 large fryer or roasting rabbit
8 or 10 medium baking or red potatoes
2 medium onions
1 lb. carrots
1 16-oz. can tomatoes
1 green pepper (optional)
4 stalks celery (optional)
¼ lb. butter or margarine
2 Tb. Worcestershire sauce
1 Tb. A-1 sauce or Heinz 57 sauce
1 t. salt
½ t. pepper
½ t. garlic salt or garlic powder
2 cups water

Cut up rabbit into serving size pieces, arrange on bottom of roast pan. Cut up carrots, onions, tomatoes, and potatoes. Place in pan; mix all other ingredients except butter. Pour over rabbit and vegetables. Divide butter into 4 or 5 pats. Place on top of vegetables. Cover. Cook at 350 degrees for 1½ hours (or at 400 degrees for 45 minutes if you are in a hurry). Serves 6. Freeze half for a busy day if desired.

Karol and Bill Lowery, Cottontail Corners Rabbitry, Seguin, Tex.

Rabbit Fricasseed in Beer

The beer adds a subtle tang

1 rabbit, cut into serving pieces
Salt and pepper
Flour
½ cup butter
2 Tb. flour
1 4-oz. can mushrooms
¾ cup beer
1 8-oz. can white onions, drained
1 cup cooked French-style green beans

Sprinkle rabbit with salt and pepper, and roll in flour. Melt butter in large skillet, add meat, and brown on all sides. Remove from pan. Blend 2 Tb. flour into remaining fat. Drain juice from canned mushrooms and add enough water to it to make 1¼ cups liquid, placing mushrooms aside. Gradually stir liquid mixture into skillet and cook, stirring constantly, until thickened. Stir in beer. Return rabbit to skillet, cover, and simmer about 1 hour, or until tender. Add mushrooms, onions and green beans, and simmer until heated through. Serves 4.

Woody and Barbara Riley, Hideaway Rabbitry, Watertown, New York

Barbecued Rabbitburger

"Our Number One Rabbit Recipe!"

2½ lbs. ground rabbit meat
2 medium onions, chopped
2 Tb. brown sugar
1½ Tb. celery salt
2 cups tomato soup (directly from can)
2 Tb. vinegar
2 Tb. prepared mustard
¾ cup catsup

Put a small amount of fat in skillet. Fry meat until fat starts to exude. Add chopped onions and fry until onion is transparent. Add remaining ingredients and simmer for ½ to 1 hour, or until thick. Serve on toasted bun.

Janet Beyer, DJ's Rabbits, Shawano, WI

Baked Rabbit in Cheese-Mushroom Sauce

An easy and delicious supper

1 frying rabbit, cut into serving pieces
Flour for dredging
2 Tb. butter, melted
1 cup light cream
1 can cream of mushroom soup
1 can cheddar cheese soup (or ½ cup shredded American cheese)
1 small jar sliced mushrooms and liquid
½ cup minced onion
Salt and pepper to taste

Coat rabbit pieces with flour. Melt butter in oblong baking dish, and add rabbit. Bake uncovered in 425 degree oven for 30 minutes. Turn meat, and bake additional 15 minutes. Remove from oven; reduce temperature to 325 degrees. Pour off excess fat from casserole dish. Combine remaining ingredients, pour over rabbit, cover and bake an additional 15 to 30 minutes, or until tender.

Rabbit Bar-B-Q

"The dish I serve most for large gatherings."

- 2 t. minced onion
- 2 Tb. flour
- 1 cup tomato sauce (8 oz.)
- 1 cup water
- 2 Tb. lemon juice
- 2 Tb. salt
- ¼ t. pepper
- 2 Tb. sugar
- 1 t. horseradish
- 1 t. mustard (dry)
- ¼ chopped green pepper
- 2 c. chopped cooked rabbit
- 1 can mushrooms

Heat some butter in pan. Add onion and cook 5 minutes. Blend flour, tomato sauce and water. Add to cooked onion in pan; heat to boiling. Add remaining ingredients, and simmer 15 minutes. Just before serving add mushrooms, and heat through. Serve on buns.

Mrs. Doris Suheski, Prairie Acres Rabbit Farm, Crystal Falls, Mich.

Rabbit Accents

Enhancing the delicate flavor of rabbit by soaking it overnight in a marinade has been a standard practice of good cooks through the centuries. Always use a large, deep bowl made of earthenware, glass, or stainless steel... never aluminum. Place your rabbit in the bowl, cover with marinade, and soak overnight or longer in the refrigerator.

Our favorite marinade recipe is:

- 2 to 2½ cups liquid (use any of the following: dry red or white wine, cider vinegar, apple cider, apple juice, chicken or beef boullion).
- 2 Tb. salad oil
- 2 t. salt
- 1 t. pepper (freshly ground is better!) or 6 black peppercorns
- ¼ t. thyme
- 1 garlic clove, crushed
- 2 onions, thinly sliced
- 1 bay leaf
- 2 Tb. parsley
- 1 Tb. sugar

Exotic additions such as ½ t. fennel seeds, or 2 Tb. crushed juniper berries, may be added.

Here are two sauce favorites that compliment the delicate flavor of rabbit.

Piquant Hunter's Sauce

- ¾ cup currant jelly
- ¼ cup orange juice
- 2 t. dry mustard
- Dash ginger
- Dash onion salt

Port Wine Sauce

Warm together ½ cup currant jelly mixed with ¼ cup port wine, ¼ cup catsup, and a dash of Worcestershire.

Bouquet Garni

A Bouquet Garni adds subtle flavoring to rabbit soups or sauces. Combine your choice of dried herbs onto a square of cheesecloth, and tie together the four ends. The Bouquet Garni may be made in advance, and stored in an airtight container. The general rule of thumb in blending herbs is: ½ t. of each herb, to 1 bay leaf.

Try combinations such as 1 bay leaf combined with parsley, green onion tops, salt and a few crushed peppercorns. Another basic combination with bay leaf is thyme, parsley, and marjoram. In some recipes you may prefer chopped parsley, 2 or 3 ribs of celery with leaves, 3 or 4 peppercorns, and a few pieces of lemon rind.

Rabbit Pate

Try this as an appetizer next time you have a party.

1 2-lb. rabbit, cut into small 1¼ pieces (or 1 2-2½ lb. frozen rabbit, thawed)
4 Tb. butter or margarine
1 clove garlic, crushed
1 onion, peeled and cut into wedges
½ cup chopped parsley
1 cup sliced mushrooms (or 1 small jar, drained)
¼ t. marjoram
¼ t. thyme
½ cup dry white wine
Puff pastry dough (frozen if desired)

Brown rabbit meat on all sides in butter. Set aside, and saute mushrooms in same skillet. Put meat and onion through food grinder, using a fine blade. Combine in skillet the garlic, parsley, mushrooms, marjoram and thyme. Add wine and blend. Roll the puff pastry dough to approximately ⅛ inch thick. Line standard load pan with the pastry, and fill with rabbit mixture. Fold pastry over top and seal tightly, making 2 or 3 small holes in dough to let steam escape. Brush with beaten egg, if desired and bake in moderate oven (350 degrees) about 1 hour. Cool and turn out of pan. Serve warm or cool. 8 servings.

A pair of well-nourished, healthy rabbits

Chapter 10

Keeping rabbits healthy

Well-fed rabbits, cared for under good conditions, are likely to avoid health problems. However, there are diseases that cause losses and you will want to recognize them and know how to deal with them.

As farmers have found with chickens, hogs, and cattle, the greater the numbers raised together, the greater the chances for disease. Pasteurellosis, ear mange and coccidiosis are the most common rabbit diseases. These infectious diseases usually are introduced by contact with infected rabbits or by mechanical carriers.

The adult rabbit is the most common carrier. When you introduce new stock, keep them isolated for 3 to 4 weeks until you are sure they are not carriers.

Growers who show rabbits expose their animals to others and are most susceptible to disease outbreaks. Rabbits at a show or fair can pick up disease from the air or from proximity to other rabbits. When returned home, show rabbits can spread disease to others. Keep them isolated for 2 weeks.

Rabbits that are carriers of disease are hard to recognize. They may appear healthy yet can shed infection in the feces, urine or droplets expelled while breathing.

The rabbit owner, unfortunately, is the major cause of mechanical transfer of disease. The grower who treats a sick animal and then moves on to check or count newborn young can be an important carrier of respiratory infection and coccidiosis.

Feed salesmen, servicemen, and pickup buyers and visitors who have made the rounds of other rabbitries may be sources of infection. The grower himself should avoid visiting other rabbitries if disease is known to be present.

These examples may be considered extreme, yet they happen repeatedly and are definite factors of introduction of disease. Dogs, cats, birds, and rodents have been incriminated in carrying diseases to rabbits and should be kept out of the rabbitry. Insects also should be controlled.

How to prevent disease problems

The way you handle your rabbits has a lot to do with their health. Most rabbit disease problems can be prevented by housing in a clean, quiet, draft-free environment. Provide clean feed and water on a regular schedule. Neglect is an open invitation to disease.

Some disease problems come with the rabbits you buy. Select only high-quality genetic stock. Look for strength and freedom from disease when you buy.

Observe every rabbit each day for any signs of illness. These signs include: dull, bluish-colored eyes; rough, lackluster hair coat; any change of behavior, posture, and movement; any change of appetite for food and water; and any change in the color, consistency, and volume of urine and feces. Remove all sick rabbits from the stock colony to a separate and isolated "sickbay" area for further observation, diagnosis, treatment, or culling.

It is usually a better practice to cull any rabbits that continue to do poorly. Too frequently, attempts are made to medicate away rabbit problems rather than humanely destroying these sick and low-quality animals. Good culling practices usually are the basis for a successful rabbitry and will result in a strong genetic strain of breeding rabbits for your stock colony.

Only use medications when there is a definite diagnosis. There are no "cure all" medications.

If you have any doubts about the diagnosis and correct treatment or the many other contributing causes to your rabbit problems, it is desirable to consult any one of the following: your local, practicing veterinarian or the state or county veterinary diagnostic laboratories. Your county agricultural agent can help you locate these services.

When medications are indicated, be sure to carefully read and accurately follow all the directions on the container label or prescription. If medicating a rabbit intended for use as meat, be sure to check the Food and Drug Administration's (FDA) recommendations for medication clearance and withdrawal requirements because most medications remain in the animal's body tissues and fluid for a period of time after administration.

Good rules for disease prevention

Sanitation is the key to disease prevention and control. Remove manure, soiled bedding, and contaminated feed every day. Clean hutches weekly. If you have some extra space, disinfect a hutch and leave it empty for a period of time. Right after weaning is a good time to do this; it protects the newborn litter.

Wash the watering and feeding equipment frequently in hot, soapy water. Rinse in clear water, drain well and place in the sun to dry. If it is the wrong season for sun, rinse in a water disinfectant solution, then rinse again in clear water.

Do a very thorough cleaning job on the nest box before each litter arrives. Feed stores carry a rabbit disinfectant that is good for this purpose.

If you do the following, you should not have much trouble with disease:

1. Be careful about lending your rabbits to others.
2. Keep new purchases of rabbits, or rabbits that you have shown at county fairs, isolated from the others for at least 3 weeks.
3. Bury or burn dead rabbits. Don't let dogs and other animals drag them around, helping to spread the disease.
4. Keep your hutch clean, and disinfect it frequently.
5. Keep your equipment clean, and prevent feed from being contaminated by manure.
6. Keep your rabbits healthy by providing enough clean food and plenty of clean, fresh water.

Guide to rabbit diseases

Diseases are caused by living organisms, such as bacteria, protozoa, fungi, and viruses. Some organisms may be present in healthy animals and not cause illness until the animal's resistance is lowered. Other organisms make the animal sick immediately. An animal's resistance can be lowered by drafts, heat, cold, overfeeding, being frightened, overhandling, sudden changes of environment and feeding practices, crowding, and poor nutrition. These are sometimes called "predisposing causes" or "stress factors."

Different diseases may cause similar symptoms or conditions. Since the unskilled eye cannot determine seriousness, losses could be severe before proper treatment is administered.

When you notice a rabbit that isn't "quite right," examine it carefully for disease or other problems. The following tables of common rabbit disease can help you determine the ailment and what to do about it.

Coccidiosis

Common and most serious disease of rabbits. Caused by an internal parasite that undergoes a complicated life cycle, part in the rabbit and part on the ground.

Symptoms: Diarrhea, progressive emaciation, dullness, and listlessness, with gradual weakness of the hind legs and finally coma or convulsions immediately before death.

Action: There is no treatment of known value for this disease. Therefore, everything should be done to keep rabbits away from feces. This is done by slatted or wire bottom hutches. These must be constructed so the fecal matter will fall below or through the floor. Hutches should be scrubbed and disinfected frequently.

Snuffles or contagious catarrh

Inflammation of mucous membrane of nasal passages and sinuses of the head.

Symptoms: Sneezing continually with a white, sticky discharge from the nose. Rubbing of nose in paws, causing hair to soil and mat. Later thinness, weakness, and death.

Action: No treatment is of any value for this trouble. Even if only one animal is ailing, it is well to regard it as a threat to the herd. Isolate the rabbit, and have another person look after it. Well-fed stock kept under strict sanitary conditions are not likely to be affected by contagious catarrh.

Common cold

Mild inflammation and discharge.

Symptoms: Often mistaken for snuffles.

Action: Isolate rabbits in dry, well-ventilated hutches, free from drafts. Sunlight and good nutritious feed and clean water should be given to the rabbits. A drop of ephedrine in each nostril will help. Chlorodyne at the rate of 3 drops in a teaspoonful of warm water is also good.

Hair swallowing or obstruction

Causes some trouble, especially in angoras. Rabbits lick the fur, and the moulting hair is taken into the mouth and swallowed. The long hair bunches and forms plugs, hereby blocking the intestine.

Symptoms: Hair may be found in the feces. The rabbit will stop eating and be listless. Constipation with intestinal fermentation will give rise to toxic symptoms, and death follows in 3 to 4 days.

Action: Drenching the rabbit with one half grain of calomel will give results. A little Epsom salts in the feed once weekly tends to keep the intestine free from accumulating hair. Best plan is to consult a veterinarian.

Necrobacillosis

Infectious disease which gains entrance to the body through small wounds, usually about the lips or face.

Symptoms: Painful swelling which spreads and may involve the face, throat, and chest regions. Wound becomes dirty, grayish-white in appearance. Sores may appear on legs and flanks. Difficulty in eating may mean rabbits may gradually starve to death.

Action: This disease does not respond to treatment. Remove the animals and thoroughly clean and disinfect all hutches, feed, and water utensils.

Pneumonia

Inflammation of the lungs, may involve one or two lungs. Occurs from exposure to drafts, sudden change in temperature, becoming wet, or specific infections. Lowering the vitality predisposes this disease.

Symptoms: Lack of appetite, breathing difficulty, droopiness, and listlessness. Watery discharge from eyes and mucous discharge from the nose.

Action: Provide well-ventilated hutches and prevent dampness and extremes of heat and cold. Since pneumonia is secondary to other infectious diseases, dispose of such animals. Deaths from pneumonia are not frequent unless there are complications.

Vent disease or spirochetosis

Affects both sexes. Venereal disease found in sores of the affected organs. It is transmitted during mating and appears as small, scaly sores on the genital organs.

Symptoms: Small, denuded places appear around the vent and slowly enlarge until small vesticles or blisters are found. These will rupture, and the area becomes covered with heavy scabs. Generally, the animals remain in good condition and exhibit no other symptoms.

Action: Examine all rabbits, male and female, before mating. Isolate all suspected animals, destroy all bedding, and thoroughly disinfect everything in and around the hutch. Do not use any suspects for breeding until they are cured. Clip the hair around the area involved. Remove discharges and crusts, but be careful not to cause bleeding. Sponge areas with cotton dipped in warm water and boric acid. The remaining scabs can be easily removed. Apply ointment of 3 parts lanolin and 1 part calomel to affected areas 3 to 4 times a week after thoroughly cleaning. Use iodine on external lesions of bucks. For interior use on the sheath, use 1:1,000 solution of potassium permananate. Lesions usually heal within 10 days or 2 weeks, but as the disease may recur, the treated animals must be examined at regular intervals.

Vent infections

Occur around the external reproductive organs and are mistaken for infections spirochetosis.

Symptoms: They occur around the vent in the form of small postules filled with a white, cheesy substance. Occasionally, one ruptures, and the spot is covered with a heavy scab.

Action: Open pustules with a sterile needle or small knife, pressing out the cheesy substance. Paint the area with iodine. Do not mate the animal until the pustules heal. This condition may be prevented by keeping the hutch floor clean, dry, and smooth.

Tularemia

Infectious disease affecting man and many species of animals. Domestic rabbits are susceptible to the disease and are used for lab tests in an effort to diagnose it, but the usual manner in which domestic rabbits are raised does not expose them to the insects that spread the disease.

Symptoms: Fever and swollen lymph nodes.

Action: Consult your veterinarian. However, it is extremely unlikely that your rabbits will ever become infected with this disease.

Eye infections

A plague for young rabbits. Organisms are picked up from the bedding material and the dirty fur of the doe.

Symptoms: A milky discharge from the eyes, causing the eyelids to be sealed together. The eyeball becomes red and appears to be covered with a film.

Action: Hutch, nest box, and feeders should be kept clean at all times. Swab the eyes with a warm, 4 percent boric acid solution, and remove the secretions 1 or 2 times daily. If the condition does not disappear in 1 week to 10 days, the use of a 5 percent solution of argyrol may bring relief. This is to be applied after cleaning the eyes with a warm boric acid solution.

Bloat or mucoid enteritis

Digestive ailment that causes great loss to the rabbit industry each year. All efforts to reduce the disease have failed, but the information has helped keep mortality at a minimum.

Symptoms: Affects rabbits of all ages, but most common among the younger animals during the first 16 months of life. First symptoms are lack of appetite, lusterless eyes that squint, rough dull fur, and grinding of teeth. Intense thirst. Later, bloated and distended abdomen. Some have diarrhea, and others are constipated.

Action: Since no effective treatment has been developed, not much can be done. Remove all feed and water for 48 hours, then give small quantities of green feed. This helps in some instances. After the fourth day, the rabbit may be given water once daily in addition to the green feed. Allow the animals to have water for only a few minutes, then take it away. After about eight days, a small quantity of good alfalfa hay may be given; a few days later, give a small portion of concentrated feed. Do not overfeed, but get the rabbits back on full feed gradually.

Sore hocks

Inflammed or ulcerous areas, either on the underside of the hind feet or on the pad and toes of the front feet. Rabbits in poor condition, either from improper feeding or overproduction, and rapidly developing young, are the most susceptible.

Symptoms: Sitting in a "cramped-up" position and refusing food. Move cautiously, and try to keep weight off feet. A sore develops on the underside of the hocks in the hind legs. It is also to be found on the front feet as well.

Action: Treatment is successful unless extensive abscesses have formed. Clip matted and adjacent fur. Wash affected areas with warm water and soap and remove all dirt and scabs. Dry all parts, apply pine tar, zinc ointment, or iodine ointment every other day until healing is well under way. Hocks will heal faster if the affected rabbits are confined to a pen on well-drained, clean sod. In such a place, mild cases of sore hocks will often heal without medication. The type of hutch floor may be a contributing factor. Unsanitary conditions on hutch floors aggravate the inflammation of the hocks.

Diarrhea

Result of infections, parasites, or other specific disease. May be caused by improper feedings.

Symptoms: Loose, watery stool, frequently offensive in odor, rapid loss of weight, and weakness. Severe diarrhea is accompanied by indigestion.

Action: Take away feed and water for 24 hours. Gradually reintroduce both until both are at the normal amounts.

Mammitis or caked udder

Develops in nursing does within 30 days after kindling, but it may also occur any time during lactation. Heavy milking does are the most susceptible.

Symptoms: Mammary glands and surrounding tissue swell. The mammary glands become red and hot. They become painful to the touch, and the does refuse to nurse the young.

Action: Reduce the grain ration radically and give green succulent feed. The mammary tissues are difficult to treat, so the affected rabbits should be placed under the care of a veterinarian.

Ear mange

Caused by a small mite which irritates the inner surface of the ear and causes the oozing of blood, which hardens and forms a crust or scab.

Symptoms: Excessive and gradually increasing moisture of the inner surface of the ear. The animal frequently shakes its head and attempts to scratch the affected area with its feet. The constant activity of the mites on the delicate skin of the ear may produce nervous symptoms, and ear mange has been ascribed as one of the causes of wry neck.

Action: Clean the ear thoroughly with slightly warmed olive oil. Remove all the caked discharge, apply a preparation composed of 2 parts iodoform, 10 parts ether, and 25 parts olive oil, petrolatum, or cottonseed oil. Another effective concoction is composed of 1 percent solution of carbolic acid or creosote in glycerine.

Chapter 11

Guide to rabbit language

Like any type of livestock or other enterprise, rabbits have a special set of words that describe parts and conditions of the animals. You'll want to refer to this section when you hear words you do not understand. Many apply to the finer points of confirmation that are important in the show ring.

Agouti—As applied to fur color in rabbits: A hair shaft possessing three or more bands of color with a definite break in color bands on entire body. Usually dark slate at base, with two or more alternating light and dark bands or rings interspersed with black guard hairs. Head, ears, and feet usually carrying ticking—belly color usually much lighter color, and not carrying ticking. For example, steel or gray Flemish, Chinchilla, and Belgian Hares.

Back—In general; the entire top portion of the animal extending from neck to tail. The area covered by the backbone or vertebrae.

Bare spots—A section of the animal entirely denuded of fur.

Bell ears—Ears which have large tips with distinct fall or lop.

Belly—The lower part of body containing the intestines—the abdomen. (For purposes of defining color area—the underbody of an animal from the forelegs to the crotch area.)

Blaze—White on the head and nose running up between the eyes on rabbits with multicolored fur.

Boils or abscess—A hard swelling or isolated collection of pus or purulent matter occurring in the rabbit's skin, accompanied by localized fever and heat.

Bowed legs—Applied to both fore and hind legs. Bent like a bow; curved outwardly in the middle.

Breed—A race or special class of domestic rabbits which reproduce distinctive characteristics of fur markings and texture; shape, size, and growth. A breed may be subdivided into varieties, as for example, the Black, Blue, and Tortoise varieties of the Dutch breed.

Breeding certificate—A written certificate by the owner of a stud buck, showing its pedigree in full and the date of breeding to a particular doe; given for the purpose of making proof of the ancestry of the young.

Broken coat—Guard hair broken or missing in spots exposing undercoat; areas where coat is affected by moult exposing undercoat.

Broken ear—A distinct break in the cartilage which prevents erect ear carriage.

Broken tail—A tail that is or has been broken and is out of line—a disqualification.

Buck—An unaltered male rabbit.

Buck teeth—Protruding or elongated teeth in either upper or lower jaw caused by malocclusion—improper alignment of the upper and lower front teeth preventing the normal wearing action.

Buff—A rich golden orange with a creamy cast.

Bull dog—As applied to head; a short, broad, bold head of pronounced masculine appearance.

Butterfly—A dark colored area on the nose such as found on the Checkered Giant and English Spot breeds.

Caked teats or caked udder—Inflammed and feverish condition of the teats arising from superabundant milk supply in a doe rabbit. Easily noted by the teats distended with hardened milk.

Carriage—The mode in which a rabbit bears itself; the style or station of the animal.

Chest—The front portion of the body between the forelegs and neck—the breast or thorax.

Cheek—The sides of the face beneath the eyes.

Choppy—As applied to type—having the back and loins cut off abruptly and falling vertically to the tail; not having a gracefully arched back and loins.

Cobby—Stout and stocky; short legged and short coupled.

Coccidiosis—Protozoa type organisms which infect the intestinal tract, resulting in poor growth and which may lead the way for secondary diseases.

Cold—A respirational infection in rabbits, characterized by sneezing; a thin watery nasal discharge, and slightly matted fur on the inside front feet.

Collar—A white band completely circling the body at the chest and forelegs.

Condition—The physical state of a rabbit in reference to health; cleanliness, texture, moult of fur, and grooming.

Conjunctivitis—(Weepy Eye) Bacterial infection of eyelids causing a discharge.

Cow hocks—Hocks that turn or bend inward causing the foot portion to turn outward.

Creamy—Light yellow; the color of cream.

Dead hairs—Lacking life—produced by molting.

Density—The property or quality of a thick coat.

Dewlap—A pendulous fold or folds of loose skin hanging from the throat.

Disqualify—One or more permanent defects, deformities or blemishes which renders an animal unfit to win an award in competition or from taking part in an exhibition. (Disqualified animals are ineligible for registration.)

Doe—A female rabbit.

Ear canker—An inflamed, swollen, scabby condition of the lower inside ear of rabbits, caused by colonies of rabbit earmites.

Ear-lacing—A black or dark-colored line of fur outlining the sides and tips of the ear.

Elimination—One or more defects presumed to be temporary and curable. Cause for elimination in a show; or from registration until cured or corrected. (Eliminated overweight or underweight animals can compete in fur class.)

Enteritis—Nonspecific and muccoid types—Causes are unknown, but considered to be of bacterial origin; stomach and intestines may fill up with water, droppings being poorly formed and appear soft and jellylike.

Eye color—As described in breed standards; the color of the iris—the circle of color surrounding the pupil.

Eye infection—Pus under eye lids in baby rabbits; eyes may stick shut caused by bacteria coupled with stress such as cold drafts.

Faking—Any change in the external appearance of a rabbit on exhibition, with intent to deceive.

Faults—Imperfections or lacking ideal type for breed and variety, not serious enough to disqualify the rabbit. Some examples are: cheek spots too large, bars instead of spots, lack of taper in spine marking, or herring bone.

Fine coat—Too fine in texture—lacking body. Guard hairs being too weak and thin in structure, similar to hairs making up the undercoat.

Flabby—The condition of a rabbit when the flesh or fur hangs loosely on the animal by its own weight—not trim and shapely.

Flank—The sides of the rabbit between the ribs and hips, above the belly.

Flat coat—Coat lying flat or close to body. Lacking spring or body as evidenced to touch. Fine coat coupled with lack of density.

Fly back—Fur properties which cause the fur to return quickly to its normal position when stroked toward the head.

Flying coat—A loose, fluffy coat of fur, caused by undue length and thinness of under wool and weak guard hairs.

Foreign color—As applied to rabbits. Any color of fur, eyes, or nails differing from the prescribed standard of perfection for the breed and variety in question.

Foot—That part of the leg on which an animal walks or stands. On the foreleg—that portion below the ankle or pastern. On the rear leg— that portion below the hock joint.

Forehead—The front part of the head between the eyes and base of ears.

Fur block—Developing does are most susceptible, yet other animals may also pull fur, consuming it in large quantities, then the rabbit goes off feed and the blockage may occur in the intestine.

Genotype—The genetic inherited characteristics and potential of the stock.

Glossy—The reflected luster or brightness from naturally healthy fur in rabbits—a natural sheen of the fur enhanced by grooming.

Guard hair—The longer coarser hair of the coat offering protection to the undercoat and furnishing wearing quality to the coat in addition to providing the sheen.

Hairline—A narrow white line running between the ears connecting the blaze and collar on the Dutch.

Heat prostration—(Heat Stroke) Rapid respiration rate, mouth open due to extremely high environmental temperature.

Hindquarters—The after portion or posterior section of the body, composed of loins, hips, hind legs and rump.

Hip—The thigh joint and large muscular first joint of the hind leg.

Hock—The middle joint or section of the hind leg between the foot and hip.

Hog fat—The condition of a rabbit obviously over-fattened, and consequently out of proportion to the true type of the breed.

Hump back—As applied to type—having a hump or protuberance on the back marring a gracefully arched outline.

Inbreeding—A system involving very close mating of stock such as brother and sister (full sib).

Inherited—The degree for which a trait or characteristic is transmitted from parent to offspring (progeny).

Inner ear—The concave portion of the ear.

Intermediate—A rabbit six months of age, or over, and under eight months of age.

Junior—A rabbit under six months of age.

Ketosis—A condition associated with kindling time when overfat does go off feed and may starve themselves to death.

Kindle—To give birth to young rabbits.

Knee—The second joint of the leg—connecting the thigh and leg. In animals—more properly called the "hock." (The second joint of the foreleg is the elbow.)

Lapin—French word for rabbit. Also, in fur trade, dyed rabbit fur.

Lazy tail—A tail that is slow to assume its normal position when moved.

Line breeding—A system of mating stock so that the progeny are highly related or close decendants of chosen ancestry.

Litter—Young of doe born at one time.

Loin—That portion of the back on either side of the spine and between the lower rib and the hip joint.

Loose coat—Fur lacking density in undercoat coupled usually with fine guard hairs and a resultant lack of texture.

Lopped ear—Pendulous; not carried erect; falling to the side or front.

Luster—Brightness and brilliance of fur.

Malocclusion—An inherited defect where upper and lower jaws do not permit the teeth to meet resulting in long, uneven teeth extending out of the mouth.

Mandolin—As applied to type, having the appearance of the body of a mandolin laid face down—back and saddle arching toward the loins to make noticeably large and broad hindquarters.

Marked—Having the basic color broken up by orderly placement on a white (usually) background.

Massive—As applied to type—bulky and heavy; ponderous; large and compact.

Mastitis—(Blue Breast) Bacterial infection causing red to purplish and/or black nipples.

Mealy—Having the appearance of being powdered or sprinkled with meal.

Meaty—The quality of being able to carry a goodly proportion of meat in proportion to the size and type of a rabbit—a noticeable meatiness at the forequarters, back, saddle, loins and haunches.

Metritis—(White Discharge) Bacterial infection of the uterus causing a white sticky discharge which may affect either or both uteri; infected does become sterile.

Molt (moult)—The act or process of shedding or changing the fur, twice yearly. The baby or nest fur is moulted at two months, and the first natural coat of fur is fully developed at four to six months.

Muzzle—The projecting portion of the head surrounding the mouth, nose and lower jaw.

Neck—That part of the animal connecting the head and body.

Nostrils—The two openings or apertures of the nose leading to the head.

Off-colored—Applied to several hairs or patches of fur foreign to the standard color of the animal.

Outcrossing—A system of mating unrelated stock of the same breed.

Pair—A male and female rabbit.

Palpation—A hand method used to determine if a doe has conceived and will bear young.

Parasites—Rabbits occasionally harbor mange, lice or fleas, but such parasites are very common.

Pasteurellosis—(Hemorrhagic Septicemia) Bacterial infection coupled with stress factors causing loss of appetite, a nasal discharge and followed by pneumonia.

Patches—A small section of fur foreign to the color standard of the animal.

Paunch—The prominent portion of the abdomen.

Pedigree—A written chart of the male and female ancestors of a rabbit, showing the date of birth; ownership of dam and the parents, grandparents and great-grandparents of the specimen in question.

Pelage—The fur coat or covering of an animal, as in rabbits.

Pepper and salt—A flat, unattractive appearance of black and white ticking, as in Chinchillas. Caused by lack of contrast and waviness, and in ticking and weakness of color in the tips of the guard hairs.

Phenotype—The appearance of the individual rabbit.

Pneumonia—Infectious bacterial organisms causing labored breathing due to congestion in the lungs.

Pot belly—A distended condition of the stomach and intestines usually found in young rabbits.

Poor coat—Fur not in good condition through moulting, rust or ill health of the animal. Also caused by inattention to grooming.

Rabbit—A domesticated rodent of the genus *Orctolagus cuniculus.*

Racy—As applied to type—slim, trim, alert and active. Slender in body and legs—harelike.

Ribs—The curved portions of the sides immediately back of the shoulders above the belly.

Ringworm—A fungus causing a loss of fur usually in circular patches on the head with red, scaly skin showing.

Rump—The hinder portion of the back and bones.

Rust—A reddish-brown coloration of fur, usually appearing on the sides, flanks or feet of rabbits, having the appearance of iron rust and being foreign to the standard color.

Saddle—The rounded, intermediate portion of the back between shoulder and loin.

Sandy—The color of sand, as in sand-gray Flemish Giants. Gray with reddish brown cast interspersed with dark guard hairs.

Screw tail—A tail that is twisted; a cork screw tail has more than one turn—a disqualification.

Self or self-colored—Animals of the same colored fur over the entire head, legs, body and tail. Not having ticking, agouti hair pattern, or shadings.

Senior—A rabbit six months of age and over in those breeds having two show room classes (Junior and Senior). A rabbit eight months of age and older in those breeds having three show room classes (Junior, Intermediate, Senior).

Shadow bars—Weakness of self-color in the fur of both fore and hind feet, appearing in the form of white or lighter colored bars running across the feet, and acting as a severe cut or penalty in scoring. Occurs more often in the agouti breeds than in selfs.

Shoulder—The uppermost joint of the foreleg, connecting it with the body.

Silvered—Having the appearance of silvery sheen or luster—an abundance of silver-white or silver tipped guard hairs interspersed through the fur so as to produce a lustrous silvery appearance.

Slipping coat—A coat that is shedding or moulting a profusion of hairs.

Slobbers—Excessive salivation creating wet, or extremely moist, and unsightly fur around the mouth and lower jaw and forelegs.

Snaky—As applied to type—slender, narrow body, thin and typical of the Himalayan.

Snipey—As applied to head—narrow and elongated, with an appearance of undue leanness.

Snuffles—A virulent contagious infection of the nasal passages and respiratory organs, usually terminating in chronic illness. Indicated by fever, heavy breathing, sneezing and discharge of thick creamy naso-pus from the nostrils.

Solid color—Of the same color uniformity over the entire animal— not mixed with any other color. Devoid of markings or shadings. Includes ticked or agouti colored animals —also includes brindled animals, but not animals of basic "Tan" pattern.

Sore hocks—An ulcerated condition of the foot pads or soles of either fore or hind feet of the rabbit. (A misnomer—not actually pertaining to the hock.)

Spraddled legs—As applied to forefeet—bowed outwardly when viewed from the front—knock-kneed. As applied to hind feet—not set parallel with the body; turned outwardly from the hock-joint.

Station—Ideal manner of standing of carriage in conformity with standard position or pose.

Stocky—Compact, stout and cobby.

Stops—In Dutch, white part of hind foot extending upward from toes ending abruptly about ⅓ way to the hocks.

Strain—A race or stock of rabbits in any standard breed of the same family blood, having the quality of reproducing marked racial characteristics.

Stringy—The quality of having ropy or sinewy finish—noticeably in the larger breeds of rabbits if not properly fattened for market.

Sway-back—As applied to type—having a distinct fall or scoop in that portion of the back between the shoulders and hind-quarters, as distinguished from gradually arching back.

Sweep—An unbroken flow of designated markings; there should be no gaps or congested areas.

Symmetry—(As applied to type in rabbits.) The quality of possessing a harmonious proportion of head, ears, legs and body structure conforming to the standard type of the breed represented.

Tail-carriage—The way in which a rabbit carries the tail. Poor tail carriage is denoted by the tail being carried to one side or the other instead of correctly.

Tattoo—A permanent method for identification of animals, a punched code in the ear.

Texture—That quality of fur pertaining to its action when stroked toward the head. The character of fur as determined by feel or touch; such as: "fine" or "coarse" in texture.

Ticking—A wavy distribution of longer guard hair throughout the fur of a color distinct from the under wool or body fur. Such ticking is usually produced by black-tipped guard hairs, as in Chinchillas, Flemish Giants and Belgian Hares, and adds greatly to the beauty of the fur.

Tucked-up—As applied to type—the trim appearance of a Belgian Hare, with long rounded body and breast and belly gathered in closely to form an arch when the animal is in a sitting position.

Type—As used in these standards—to denote body conformation of a rabbit, or shape of a particular part of the animal as in "Head type". The general description of the physical makeup of the animal.

Typical—Serving as an ideal representative of any given breed or variety as applied to type, color, or fur quality.

Under-color—The base of the fur hair—shaft next to the skin, not the belly fur of the animal.

Urine or hutch burn—Bacterial infection of external sex organs and anus—to avoid, use all wire cages.

Variety—(As applied to rabbits, "Type shows the breed and color the variety.") A breed subdivision applicable to animals of any recognized standard breed, but distinct in color of fur from other races or subdivisions thereof.

Vent disease—Venereal diseases in rabbits of both sexes. Indicated by scabby, reddened male or female organs, usually exuding pus.

Wall eyes (moon eyes)—Having a milky film over the cornea, or appearance similar to a moonstone. Colored eyes having an extremely light iris giving a glazed appearance.

Wool—Applied as descriptive of the fur of Angora rabbits—the guard hairs and underfur being from 2½ to 5 inches in length and resembling fine wool in texture.

Wolf teeth—Protruding or elongated teeth in either upper or lower jaw caused by malocclusion—improper alignment of the upper and lower front teeth preventing the normal wearing action.

Wry-tail—Abnormally bent, curled or twisted permanently held to one side—a disqualification.

Where to get more information

There are magazines and publications available to provide more information about rabbit production and to keep you up to date. Here are some of the materials available. Write to them for information on costs.

Magazines

Countryside & Small Stock Journal
Editor-Publisher, Jerome Belanger
Rt. 1, Box 239
Waterloo, WI 53594

Domestic Rabbit-American
Rabbit Breeders Association
1925 S. Main St., Box 426
Bloomington, IL 61701

The Journal of Applied Rabbit Research
Oregon State University
Corvallis, OR 97331

Rabbits Magazine (monthly)
Editor, Sue Pfrang
Hwy. 19, East
Waterloo, WI 53594

The Rabbit and Cavy Gazette
Editor, Charles Clark
1725 S.W. Blvd.
Kansas City, KS 66103

The Small Stock Farmer Magazine
Eastern R&CFA
750 Grant St.
Hazelton, PA 18201

Books and Circulars

Rabbit Book
Ralston Purina Co.
Checkerboard Square
St. Louis, MO 63188 (20 pages, 1981)

Rabbits
Carnation Milling Division
1800 Potter Ave.
Kansas City, KS 64126 (40 pages, 1981)

Raising Rabbits
Ann Kanable, Rodale Press
Emmaus, PA 18049

Commercial Rabbit
Market Directory
P.O. Box 35855
Houston, TX 77035

Correspondence Course

Penn State University
Rabbit Correspondence Course
202 Agricultural Education Bldg.
University Park, PA 16802